STUDENT WORKBOO

ELEMENTAL GEOSYSTEMS

A FOUNDATION IN PHYSICAL GEOGRAPHY

ROBERT W. CHRISTOPHERSON

Prentice-Hall, Inc., Englewood Cliffs, N.J. 07632

Production Editor: *Joan Eurell*
Supplement Acquisitions Editor: *Wendy Rivers*
Production Coordinator: *Julia Meehan*

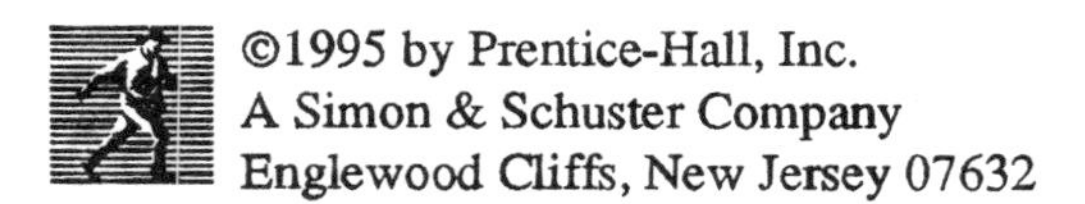

Printed in the United States of America

10 9 8 7 6 5 4 3 2 1

ISBN 0-13-362807-8

Prentice-Hall International (UK) Limited, *London*
Prentice-Hall of Australia Pty. Limited, *Sydney*
Prentice-Hall Canada Inc., *Toronto*
Prentice-Hall Hispanoamericana, S.A., *Mexico*
Prentice-Hall of India Private Limited, *New Delhi*
Prentice-Hall of Japan, Inc., *Tokyo*
Simon & Schuster Asia Pte. Ltd., *Singapore*
Editora Prentice-Hall do Brasil, Ltda., *Rio de Janeiro*

Elemental Geosystems
A Foundation in Physical Geography

Geosystems Workbook

Contents

Introduction

Outline

- On Being a Student
    - Getting Organized
    - Study Methods and the Text
- Critical Thinking Essentials
- Format for this Study Guide
- A Tour of ELEMENTAL GEOSYSTEMS
- Ready Reference
    - Table of Contents
    - List of FYI Reports
    - List of News Reports
- Map Exercise
- Geography I.D.

Welcome to a foundation in physical geography! Thus begins ELEMENTAL GEOSYSTEMS–A FOUNDATION IN PHYSICAL GEOGRAPHY. The success of the Geosystems approach throughout the United States and Canada is owed to the students and teachers who have embraced our approach to physical geography. Therefore ELEMENTAL GEOSYSTEMS is dedicated to you the students and your teachers. *"To all the students and teachers of Earth, our home planet, and its sustainable future."*

Physical geography is a basic introductory science course and as such is an opportunity to satisfy a general educational requirement in a timely and interesting Earth class. As the last sentence in the Preface states:

> **Physical geography teaches us about the intricate supporting web that is Earth's environment. Dramatic changes that demand our attention and understanding are occurring in many Earth-human relationships as we approach the new millennium. All things considered, this is an important time to be enrolled in a relevant geography course!**

Physical geography is an excellent introduction and perhaps your only exposure to science and scientific methodology (see the FYI Report "Scientific Method" in Chapter 2, page 39). Because of the nature of physical geography, a diverse and dynamic approach is at the heart of the text.

Whatever you thought geography was going to be when you went through the catalog, schedule, and registration, remember that all of us are geographers at one time or another. Our mobility is such that we cover great distances every day. Physical systems are expressed in complex patterns across Earth. Understanding both our lives and the planet's systems requires the tools of spatial analysis, which are at the core of geography.

The goal of physical geography is to explain the spatial dimension of Earth's natural systems—its energy, air, water, weather, climates, landforms, soils, plants, and life forms. For information about geography as a career possibility contact the Association of American Geographers (AAG's address is in Appendix A of the text) and ask for their pamphlets, "Careers in Geography" by Salvatore J. Natoli (1990), "Geography: Today's Career for Tomorrow," and "Why Geography?"

Earth is a place of great physical and cultural diversity yet people generally know little about it. Recent headlines have warned, "We Are Geographically Ignorant," or "Environmental Concerns Get Lost on the Map!" As we approach the new millennium, critical developments concerning human-environment themes and are already under way. Geography, as a spatial human-Earth science, is in a unique position among sciences to synthesize and integrate the great physical and cultural diversity facing us. Only through relevant education can an informed citizenry learn about the life-sustaining environment that surrounds our lives. To get started complete the "Geography I. D." on page 8 of this introduction.

The purpose of this introduction is to give you some strategies for getting an *A* in this physical geography course and, ideally, to better perceive your relationship with Earth. Not bad for an initial goal! Also, a few hints are presented to make the overall learning process easier and possibly improve your success in all your classes. Juggling lecture and lab classes, study time, a job, family, and social activities is probably one of the toughest tasks in life; and all are perhaps happening at a time when you may be dealing with the realization that these are the major leagues of life, the time and place for foundations to be built, and for careers and a meaningful future to be started.

On Being a Student

Teacher and student share common goals in education—thinking and learning. Teachers have taken on a life-time association with the classroom whereas the student passes through the classroom and on to a degree and a career. Many people think the goal of college is focused on grades, GPA, scholarship competition, and future job placement. Certainly and undeniably, these are important institutional goals and the immediate reward for doing well in classes. Ideally however, the reward for being a student is thinking and learning. Thinking is the operation of the mind and a function of the brain, and learning represents a change in behavior.

Straight knowledge acquisition is at the simplest level of cognitive activity; applying that knowledge toward real action and behavior change is true learning. What if a student gets an *A* on a water conservation test and then goes home and wastes water? Knowledge acquisition occurred but not learning.

Thinking and learning, not grades, are the rewards for being a student. Grades are important and at the heart of institutional progress but they are not the ultimate reward for being in school.

What about the teacher? How does thi statement of setting idealistic goals apply t teaching about water conservation and the going home and wasting water? We can re state the axiom as follows: *thinking an learning, not pay, are the rewards for teach ing.* Certainly pay is important for it is a ca reer and represents the teacher's livelihood But pay is merely the institutional reason fo participating. We, teacher and student, ar linked in this critical educational pursuit.

The reason for going into this in you study guide is to give you some insight into th educational philosophy of the author of the text book you are about to read.

Getting Organized. A major aspect of you thinking and learning goals should be orga nization. Work out a detailed weekly sched ule by hour for classes, study, work, social ac tivities, sleeping, exercise, etc. Establish routine around these activities, for time man agement is going to be a major organizationa challenge. Find a comfortable study place o campus such as a specific carrel or table in th library.

For study at home, I recommend that stu dents build a desk if they do not already hav one. A plywood board and cinder (masonry blocks or chimney tiles, placed against a wall and you're ready to go for under $20. Having desk is extremely important to your college ca reer, and quite superior to the bed, dining room table, floor, or couch in front of the television Your desk is an excellent signal to others i your life that you are serious about college an career building. Now, birthdays and gift day can be focused on this pursuit with equipmen for classes: a desk pad, stapler, paper clips pencils, pens, colored pencils, a calculator protractor, metric ruler, manila folders, a col legiate dictionary, *Roget's Thesaurus* (an es sential word finder), Strunk and White' *Elements of Style* (a guide to elementar grammar, word usage, and style), paper binders, Post-It Notes, tab dividers, etc.

Develop a system for organizing you binder, handouts, and notes; experiment wit a system until you find one with which you are comfortable. I have seen bright students fall to poor grades by arriving in class with piles of disorganized papers. For each class, set up a method of coordinating reading notes with lecture notes. You may want to draw a dividing line on each sheet of binder paper about 7.5 cm (3 in.) in from the edge of the page. Reading notes can be placed on the left side and lecture notes on the right side of this line. The line can be adjusted for each class depending on the teacher's presentation.

Study Methods and the Text. For your class notes, develop a personal style of outlining, hierarchical paraphrasing, succinct ways of recording lecture and reading notes. Be attentive to your teachers for often they will tell you what items are important to know. I have selected 100 figures from the text along with some of the critical integrative tables for overhead transparencies that your instructor may use. As these are used in lecture, your teacher will communicate which material to emphasize.

There is a range of opinion about the use of highlighting pens and the marking of sentence after sentence in each chapter. I'm not convinced that this works when compared with writing out succinct reading notes that correlate to your class notes. Working the content through your mind, translating it into your own words, and recording it in your handwriting (or keyboarding) impresses the material on your mind. The more conscious the process, the better the result. Underlining another person's words may not involve conscious mental activity on your part.

I have taken care in writing ELEMENTAL GEOSYSTEMS to consider the reader. Five hundred and fifty **boldfaced** terms and concepts are presented with their definitions the first time they occur in the text. They subsequently are used in the context of that definition. The glossary presents definitions that are keyed to the way the terms are used in the text. I recommend the use of flash cards with the terms or processes on one side and a sentence or two defining and using the term on the other side. These cards can be used anywhere: at the mirror in the morning, at the bus stop, during meals, or at traffic lights.

ELEMENTAL GEOSYSTEMS is organized in a way to assist the reader with a logical flow of topics. Subjects are presented in the sequence in which they occur in nature, or in a manner consistent with history and the flow of events. I hope this systems organization will assist you with your reading. The entire text is organized around three orders of headings to give you a hierarchical nesting of subjects. The first and second order headings are printed at the beginning of every chapter, and the key terms and concept (boldfaced) words are listed at the end of the chapter.

Also note the study questions at each chapter's end; they are derived directly from the text. Suggested readings are listed for each chapter in the back of the book. I selected these readings, not only because they represent important source materials for further reading, but also because they are readily available, including journals and magazines. This text should provide you with many ideas for term paper and writing assignments that may arise in other classes. To help you further, I include a list of agencies, organizations, and reference materials in Appendix A.

The text and all figures are in metric units and the International System of Units (SI) where appropriate. Canada uses the metric system exclusively, whereas English measurement equivalencies are presented for the United States which is still in a transition period. A complete set of measurement conversions is conveniently presented in an easy-to-use arrangement inside the back cover of the text. For more information about metric conversion materials contact: the Metric Program Office, National Institute of Standards and Technology, U.S. Department of Commerce, Gaithersburg, MD 20899. A useful report was prepared by Barry N. Taylor, ed., *Interpretation of the SI for the United States and Metric Conversion Policy for Federal Agencies*, NIST Special Publication 814, NIST, Department of Commerce, October 1991.

Critical Thinking Essentials

"Critical thinking" involves the deliberate and conscious control of the thinking process by the thinker. This literally means that you think about the stages of your thinking process. Simple attention to this concept will make the classroom process leap to exciting levels.

Do not confuse critical thinking practices with course or textbook content, rather it should be thought of as the organization of and approach to content and the tactics and strategies for learning the material. You should approach the content of a course in a way that helps you learn to reason and see the application of the material to your life. Learning is defined as a change in behavior in the learner, not merely knowledge acquisition (memorization). Critical thinking effectively leads to such personal change and improvement. Critical thinking assesses and improves itself continually—so the practice of it makes it grow in you!

As you learn, you need to ask: What do I want to accomplish? How can I measure if I am accomplishing this goal? What are my questions? What do I need to answer them? What other points of view are available to me? Why did I select the point of view I now have? Can I justify my answers or have I missed something or taken something for granted? How clearly am I expressing through speech and writing my ideas and findings? Can I still relate my path and direction of travel (and aspirations) to my goal? Where am I right now? How does this position relate to my beginning goal statement?

In this workbook you are asked in each chapter to select any five of the learning objectives and work through a form of critical thinking. You are asked to consciously "walk through" your learning process with these assessment questions:

- What did you know about the objective before you began?
- What was your plan to complete the objective?
- Which information source did you use i[n] your learning (text, or other)?
- Were you able to complete the action stated in the objective? What did you learn?
- Are there any aspects of the objective about which you want to know more?

Critical thinking requires that we set goal[s] for all our courses at the beginning of the term[:] purpose of each class, its importance to ou[r] lives, the method of assessment used and wh[y] this method is part of the purpose, a statement [of] the questions we want to answer, and the leve[l] of emotional, mental, and social energy th[e] course will tap. This deliberate approach wi[ll] signal a lot to you and put the term on a highe[r] academic plane—*one that involves most of yo[u] in the process.*

For more on critical thinking see: Richar[d] Paul, *Critical Thinking–How to Prepar[e] Students for a Rapidly Changing World[,]* Santa Rosa, CA: Foundation for Critica[l] Thinking, 1993; and, Linda Elder an[d] Richard Paul, "Critical Thinking: Why W[e] Must Transform Our Teaching," *Journal o[f] Developmental Education* 18, no. 1 (Fall 1994[):] 34-5; and the publication *Educational Vision[:] The Magazine for Critical Thinking*, pub[-]lished 4 times a year by The Foundation fo[r] Critical Thinking, 4655 Sonoma Mountai[n] Road, Santa Rosa, CA 95404, 1-707-546-0629.

Format for this Workbook

Each chapter of this workbook matches th[e] same numbered chapter in ELEMENTA[RY] GEOSYSTEMS. The format utilized for eac[h] study chapter is in the following sequence:

Chapter Overview
Learning Objectives
Outline Headings and Glossary Review (Key terms in the same order as presented in the textbook)
Learning Activities and Critical Thinking
Sample Self-test (Answers appear at the end of this workbook.)

I recommend that you first preview each chapter of the textbook: look at the opening photograph, read the headings and the opening paragraph, leaf through and examine the figures and captions, sample topic sentences, read the chapter summary, and look through the review questions and suggested readings. Next, read the overview and learning objectives from this study guide. As you read through the chapter, record in your own words the definition/application of each boldfaced term in your notes. This workbook provides a list of all the headings in each chapter and the boldfaced key terms as they occur under those headings. I include a small check-off box to help you keep track of your progress.

Save the learning activities and exercises until after you have read through the text chapter the first time. End with the review questions at the end of the chapter. You may want to sketch answers in your reading notes for those review questions. Finish your work by taking the self-test in this study guide.

A Tour of ELEMENTAL GEOSYSTEMS

Earth's physical geography systems are complex, interwoven threads of energy, air, water, weather, climates, landforms, soils, plants, animals, and the physical Earth itself. ELEMENTAL GEOSYSTEMS is thoroughly up-to-date, containing the latest information about the status of Earth's physical systems as viewed through the *spatial analysis* approach. ELEMENTAL GEOSYSTEMS is an introductory text in which relevant human-environmental themes are integrated with the principle core topics usually taught in a course of physical geography. The text is conveniently organized into four parts that logically group chapters with related content.

Particular attention was invested in the writing this text to achieve readability and clarity appropriate to introductory university- and college-level courses. A complete list of text features appears in the Preface of the book; however, some important highlights that you may find helpful include the following.

- Fourteen **"FYI Report"** inserts, many with figures, highlight in greater depth key topics related to text material. The emphasis varies among the focus studies from traditional (Chapters 1, 7, and 14) to topical subjects (Chapters 2, 5, 9, 12, and 16), with some handling specific tools as in the focus studies in Chapters 2 and 3. See the complete list of focus studies for the text at the end of this section.
- Fifty **"NEWS Reports"** throughout the text present late-breaking environmental issues, new applications in geography, and fascinating items that highlight content. A sample of titles includes: "GPS: A Personal Locator," "Careers in GIS," "New UV Index Announced to Help Save Your skin," "Harvesting Fog," "Water in the Middle East: Running on Empty," "Niagara Falls Closed for Inspection," "Engineers Nourish A Beach," "Worldwide Coral Bleaching Worsens," "Humans Dump Carbon into the Atmosphere," and "Huge Oil Spills." See the complete list of news reports for the text at the end of this section.
- *Elemental Geosystems* ends with a capstone chapter, "Earth, Humans, and the New Millennium," which is unique in physical geography. The text attempts to look holistically at Earth and the world in this concluding treatment. This chapter can be used to facilitate class discussions concerning the impact of humans on Earth systems, as well as the important role played by geographic analysis and physical geography in particular. An FYI Report detailing the 1992 Earth Summit is presented in Chapter 17.
- Appendix A contains addresses of important geographic and environmental organizations, general reference works, and agencies.
- Appendix B presents "Topographic Map Symbols–USGS."
- You will notice the inclusion of the latest relevant scientific information throughout the text. This helps you see physical geography as an important Earth system

science linked to other sciences in this era of expanded data gathering and computer modeling. Geography is a dynamic science not a static one! As an example, during October 1994 the Space Shuttle *Endeavour*, using radar and other remote sensors spent two weeks in Earth orbit studying many aspects of our planet's physical geography.

- An up-to-date treatment of climate change, including global warming, is integrated throughout the text. You are shown how climate change specifically relates to many aspects of physical geography. Note that the topic does not appear as one of the FYI Reports; rather, it is treated within the text. Although it is politically controversial there is a level of scientific consensus relative to global warming, with this difference clearly explained. The four reports from the Intergovernmental Panel on Climate Change (1990 and 1992) provide an important resource for these sections. During 1994, a fifth report from the IPCC was released and it confirmed and strengthened the earlier consensus among scientists that humans are causing a warming of global temperatures through the release of greenhouse gases.
- In key chapters (climate, arid landscapes, soils, biomes), large, integrative tables are presented to help synthesize content. If you choose, the rest of the chapter can be studied for deeper analysis of the subject. This approach adds to the adaptability of the text for different levels of presentation and for quarter, trimester, or semester length classes. In fact, Table 16-2 in Chapter 16 can be referenced throughout the term, as the total Earth system is organized within the various terrestrial biomes.
- Six topographic maps are presented at the back of the text to illustrate various landscapes. These are:
  - **Map #1**: Floodplain and fluvial processes (Philip, MS)
  - **Map #2**: Oxbow lake, political boundary (Carter Lake, NE)
  - **Map #3**: Folded mountains, water gap (Cumberland, MD)
  - **Map #4**: Alpine glacier features (Mount Rainier, WA)
  - **Map #5**: Landscape of continental glaciation (Jackson, MI)
  - **Map #6**: Coastal erosional and depositional processes (Point Reyes, CA)
- A world map of the Köppen-Geiger climate classification is conveniently located inside the front cover.

For Ready Reference

The "Table of Contents" and a complete listing of "FYI Reports" and "News Reports" is printed here for convenient reference. Check-off boxes are included if you choose to track your progress through the text.

Table of Contents

Chapter 10- Weathering, Karst Landscapes, and Mass Movement ❑
Chapter 11- River and Related Landforms ❑
Chapter 12- Wind Processes and Desert Landscapes ❑
Chapter 13- Coastal Processes and Landforms ❑
Chapter 14- Glacial and Periglacial Landforms ❑

PART FOUR: Biogeography

Chapter 15- The Geography of Soils ❑
Chapter 16- Ecosystems and Biomes ❑
Chapter 17- Earth, Humans, and the New Millennium ❑

FYI Report List

Chapter 1
❑ 1-1: Measuring Earth in 247 B.C.
❑ 1-2: The Search for Longitude
Chapter 2
❑ 2-1: The Scientific Method
❑ 2-2: Stratospheric Ozone Losses: A Worldwide Health Hazard
Chapter 3
❑ 3-1: Solar Energy Collection and Concentration
Chapter 3
❑ 3-2: Temperature Concepts, Terms, and Measurements
Chapter 4
❑ 4-1: Wind Power: An Energy Resource
Chapter 5
❑ 5-1: 1992—A Year of Hurricanes
Chapter 6
❑ 6-1: Ogallala Aquifer Overdraft
❑ 6-2: Acid Deposition: A Blight on the Landscape
Chapter 7
❑ 7-1: The El Niño Phenomenon
Chapter 8- – –
Chapter 9
❑ 9-1: The 1980 Eruption of Mount Saint Helens
Chapter 10- – –
Chapter 11
❑ 11-1:Floodplain Management Strategies
Chapter 12
❑ 12-1:The Colorado River: A System Out of Balance
Chapter 13
❑ 13-1:An Environmental Approach to Shoreline Planning
Chapter 14
❑ 14-1:Deciphering Past Climates: Paleoclimatology
Chapter 15- – –
Chapter 16
❑ 16-1:Biodiversity and Biosphere Reserves
Chapter 17
❑ 17-1:Earth Summit 1992

News Report List

Chapter 1
❑ #1: "GPS: A Personal Locator"
❑ #2: "A New Clock for UTC"
❑ #3: "Careers in GIS"
Chapter 2
❑ #1: "Scientists Discover the Atmosphere's Origin"
❑ #2: "Does Solar Wind Affect Earth's Weather Patterns?"
❑ #3: "Earth's Atmosphere Unique Among Planets"
❑ #4: "New UV Index Announced to Help Save Your Skin"
❑ #5: "Air Pollution Abatement–Costs vs. Benefits"
Chapter 3 none
Chapter 4
❑ #1: "Coriolis, A Forceful Effect on Drains?"

- ❑ #2: "Winds Can Be 'foolish' and 'uncertain'"
- ❑ #3: "Jet Streams Affect Flight Times"

Chapter 5

- ❑ #1: "Breaking Roads and Pipes and Sinking Ships"
- ❑ #2: "Harvesting Fog"
- ❑ #3: "Mountains Set Precipitation Records"
- ❑ #4: "Recent Tornado Records Are a Call for Action"
- ❑ #5: "Benefits from Hurricanes?"

Chapter 6

- ❑ #1: "Water in the Middle East: Running on Empty"
- ❑ #2: "Personal Water Use"

Chapter 7

- ❑ #1: "*Cwa* Climate Region Sets Precipitation Records"

Chapter 8

- ❑ #1: "Drilling the Crust to Record Depths"
- ❑ #2: "Canyon Rocks Harder than Steel"
- ❑ #3: "Yellowstone on the Move"

Chapter 9

- ❑ #1: "James Michener On Terranes and Tectonics in *Alaska*"
- ❑ #2: "Damage Strikes at Distance from Epicenters"

Chapter 10

- ❑ #1: "Alabama Hills—A Popular Movie Location"
- ❑ #2: "Amateurs Make Cave Discoveries"
- ❑ #3: "Debris Avalanche Destroys Colombian Town"

Chapter 11

- ❑ #1: "Climate Change and a River Basin"
- ❑ #2: "Niagara Falls Closed for Inspection"
- ❑ #3: "The 1993 Midwest Floods"

Chapter 12

- ❑ #1: "War Tears up the Pavement"
- ❑ #2: "Yardangs from Mars"
- ❑ #3: "The Dust Bowl"

Chapter 13

- ❑ #1: "Sea Level Varies Along the U.S. Coastline"
- ❑ #2: "Killer Waves Strike Beachcombers"
- ❑ #3: "Warmer Water Raises Sea Level"
- ❑ #4: "Engineers Nourish A Beach"
- ❑ #5: "Hurricane Hugo Devastates the Beachfront"
- ❑ #6: "Worldwide Coral Bleaching Worsens"

Chapter 14

- ❑ #1: "South Cascade Glacier Loses Mass"
- ❑ #2: "Glacial Erratic Marks Famous Grave"
- ❑ #3: "The Great Salt Lake Floods"

Chapter 15

- ❑ #1: "Soil Losses Estimated"
- ❑ #2: "Death of the Kesterson National Wildlife Refuge"

Chapter 16

- ❑ #1: "Humans Dump Carbon into the Atmosphere"
- ❑ #2: "Yellowstone Fire Assessment and Fire Ecology"
- ❑ #3: "Large Marine Ecosystems: A Management Tool"
- ❑ #4: "Annual Losses to the Rain Forest Estimated"

Chapter 17

- ❑ #1: "Huge Oil Spills"
- ❑ #2: "ANWR Faces Threats"

Map Assignment

In Chapter 3 of the text (p. 94) the following suggestion appears:

> **With each map, begin by finding your own city or town and noting the temperatures indicated by the isotherms (the small scale of these maps will permit only a general determination). Record the information from these maps in your notebook. As you work through the different maps throughout this text, note atmospheric pressure and winds, precipitation, climate, landforms, soil orders, vegetation, and terrestrial biomes. By the end of the course you will have recorded a complete profile of your specific locale.**

I recommend that you follow through on this suggestion and maybe even establish a special section in your notebook. You might want to expand your entries to include your state or province and the general region where you live.

The "Geography I. D." Assignment:

(* Derived from *Applied Physical Geography: Geosystems in the Laboratory* by Robert W. Christopherson, © 1994 Macmillan College Publishing Company.)

GEOGRAPHY I.D.

* **To begin:** complete your personal *Geography I. D.* Use the maps in your text, an atlas, college catalog, and additional library materials if needed. Your instructor will help you find additional source materials for data pertaining to the campus.

On each map find the information requested noting the January and July temperatures indicated by the isotherms (the small scale of these maps will permit only a general determination), January and July pressures indicated by isobars, annual precipitation indicated by isohyets, climatic region, landform class, soil order, and ideal terrestrial biome. Record the information from these maps in the spaces provided. The completed page will give you a relevant geographic profile of your immediate environment. As you progress through your physical geography class the full meaning of these descriptions will unfold. This page might be one you will want to keep for future reference.

GEOGRAPHY I. D.

NAME:______________________________ CLASS SECTION:________________________

HOME TOWN:____________________ LATITUDE:_________ LONGITUDE:________

COLLEGE/UNIVERSITY:____

CITY/TOWN:_________________________ COUNTY (PARRISH):___________________

LOCATION: STANDARD TIME ZONE:______________________________________.

LATITUDE:____________________LONGITUDE:__________________________

ELEVATION (include location of measurement):____________________________

PLACE (tangible and intangible aspects that make this place unique):_____________

____.

REGION (aspects of unity shared with the surrounding area):____________________

____.

POPULATION:

CITY: ____.

METROPOLITAN STATISTICAL AREA (CMSA, PMSA, if applicable):________________

_________________________________.

ENVIRONMENTAL DATA:

JANUARY AVG. TEMPERATURE:________JULY AVG. TEMPERATURE:___________

JANUARY AVG. PRESSURE (mb): _______JULY AVG. PRESSURE:________________

AVERAGE ANNUAL PRECIPITATION (cm/in.):__________________________________

AVG. ANN. POTENTIAL EVAPOTRANSPIRATION (if available; cm/in.):___________

CLIMATE REGION (Köppen Symbol and name description):____________________

____.

TOPOGRAPHIC REGION OR STRUCTURAL REGION:_______________________________

____.

BIOME (terrestrial ecosystems description): ___________________________________

____.

Name:______________________________ Class Section:_______________

Date:__________________ Score/Grade:____________

Foundations of Geography

1

Chapter Overview

"Foundations of Geography" contains the basic tools for you to use in studying the content of physical geography: spatial analysis, systems methodologies, models, the physical planet, geographic grid coordinates, time, cartography and map making, topographic maps, remote sensing, and geographic information systems (GIS). With the completion of this chapter you have the foundation necessary to work through the remaining chapters of this text. The complete outline headings for this chapter are:

Learning Objectives

The following learning objectives help guide your reading and comprehension efforts. The operative word is in *italics*. Read and work with these carefully and note that exercise #1 asks you about five of these objectives. After reading the chapter and using this workbook, you should be able to:

1. *Define* geography and physical geography in particular.
2. *Describe* what is meant by the term spatial analysis and *list* words that denote geographic content.
3. *Explain* the difference between geography and other academic disciplines.
4. *Identify* and *compare* each of the five geographic themes as established by the Association of American Geographers and the National Council for Geographic Education.
5. *Construct* the geographic continuum in a simple sketch and *identify* the relationship among subdisciplines within physical geography and cultural–human geography and related academic disciplines.
6. *Relate* the geographic continuum to the basic duality in humans and their relationship to Earth's systems.
7. *Define* an open and a closed system and *use* each in a sentence applicable to Earth's systems.
8. *Differentiate* between abiotic and biotic systems and *describe* Earth's four spheres as models of these systems.
9. *Illustrate* Earth's diameter, circumference, and shape in a simple sketch.
10. *Define* latitude and parallel, longitude and meridian, and *use* them in a simple sketch to *demonstrate* how Earth's reference grid is established.
11. *Differentiate* among Earth's principal latitudinal geographic zones.
12. *Relate* the time where you are with world standard time and *describe* the basis for UTC time determinations.

13. *Contrast* the prime meridian with the International Date Line.
14. *Explain* the relationship of cartography to both mechanical sciences and aesthetic practices.
15. *Describe* map scale and *give* an example of large, medium, and small scales.
16. *Define* map projection and *discuss* its essential properties.
17. *List* the positive and negative aspects of the Mercator projection.
18. *Identify* the types of map projections used in ELEMENTAL GEOSYSTEMS and *give* some specific examples.
19. *Describe* a topographic map and *relate* its principal uses.
20. *Portray* the National Mapping Program in the United States.
21. *Define* remote sensing, both passive and active sensing systems, and *identify* several among the 70 remote-sensing images in ELEMENTAL GEOSYSTEMS.
22. *Define* a geographic information system (GIS) and *explain* the potential of this methodology for environmental planning, assessment, and problem solving.

Outline Headings and Glossary Review

These are the first- second-, and third-order headings that divide this chapter. The key terms and concepts that appear **boldface** in the text are listed under their appropriate heading in bold italics; these highlighted terms appear in the text glossary. A check-off box is placed next to each key term so you can mark your progress through the chapter as you define these in your reading notes or prepare note cards.

The Science of Geography

- ❑ ***geography***
- ❑ ***location***
- ❑ ***place***
- ❑ ***movement***
- ❑ ***region***
- ❑ ***human-Earth relationships***
- ❑ ***spatial analysis***
- ❑ ***process***
- ❑ ***physical geography***

The Geographic Continuum

Earth's Systems Concepts

Systems Theory

- ❑ ***system***
- ❑ ***open system***
- ❑ ***closed system***

Example

Feedback

- ❑ ***feedback loops***
- ❑ ***positive feedback***
- ❑ ***negative feedback***

Equilibrium

- ❑ ***equilibrium***
- ❑ ***steady-state equilibrium***

Models of Systems

- ❑ ***model***

Earth As A System

Earth's Energy Equilibrium: An Open System

Earth's Physical Matter: A Closed System

Earth's Four Spheres

- ❑ ***abiotic***
- ❑ ***biotic***
- ❑ ***atmosphere***
- ❑ ***hydrosphere***
- ❑ ***lithosphere***
- ❑ ***biosphere***

Atmosphere

Hydrosphere

Lithosphere

Biosphere

- ❑ ***ecosphere***

A Spherical Planet

- ❑ ***geodesy***
- ❑ ***geoid***

Location on Earth

Latitude

- ❑ ***latitude***
- ❑ ***parallel***

Latitudinal Geographic Zones

Longitude

- ❑ ***longitude***
- ❑ ***meridian***
- ❑ ***prime meridian***

Great Circles and Small Circles

- ❑ ***great circle***
- ❑ ***small circle***

Prime Meridian and Standard Time

- ❑ ***Greenwich Mean Time (GMT)***

International Date Line

- ❑ ***International Date Line***

Coordinated Universal Time

- ❑ ***Coordinated Universal Time***

Daylight Saving Time

- ❑ ***Daylight Saving Time***

Maps, Scales, and Projections

- ❑ ***cartography***

Map Scales

- ❑ ***scale***

Map Projections

- ❑ ***map projection***

Properties of Projections

- ❑ ***equal area***
- ❑ ***true shape***

The Nature and Classes of Projections

- ❑ ***Mercator projection***
- ❑ ***rhumb line***

Maps Used in this Text

- ❑ ***Goode's homolosine*** ❑ ***projection***
- ❑ ***Robinson projection***

Mapping and Topographic Maps

- ❑ ***planimetric map***
- ❑ ***topographic map***
- ❑ ***contour line***

Remote Sensing and GIS

Remote Sensing

- ❑ ***remote sensing***

Active Systems

Passive Systems

Geographic Information System (GIS)

- ❑ ***Geographic Information System (GIS)***

SUMMARY

Learning Activities and Critical Thinking

1. Select any five learning objectives from the list presented at the beginning of this chapter. Place the number selected in the space provided (no need to rewrite each objective). Using the following questions as guidelines only, briefly discuss your treatment of the objective.

- What did you know about the objective before you began?
- What was your plan to complete the objective?
- Which information source did you use in your learning (text, or other)?
- Were you able to complete the action stated in the objective? What did you learn?
- Are there any aspects of the objective about which you want to know more?

a)_____:____

____

____

____

____.

b)____:____

____

____

____

____.

c)____:____

____

____

____

____.

d)____:____

____

____

____

____.

e)____:____

____

____

____

____.

2. Using Figure 1-1, p. 3, analyze and describe why you think the five photographs were selected to represent each geographic theme.

a) location:____

____

____

____

____.

b) place:

c) movement:

d) regions:

e) human-Earth relationships:

3. List, in your own words, the eight critical environmental concerns given on page 5 of the text. Following discussion with others, add any two additional concerns (geography, environment, etc.) that you think should be considered. State your reasons for us to consider them.

a)

b)

c)

d)

e)

f)

g)____.

h)____.

i) Additional concern:____

____

____.

j) Additional concern:____

____

____.

4. Inside the boxes provided (below and on the next page), <u>sketch</u> a simple open system and a simple closed system as in Figures 1-3 and 1-4. <u>Indicate</u> flows of energy and matter into and out of the system and within the system. Identify the components of each system. Your sketch may be schematic (like 1-3) or be an actual depiction of a real system (like 1-4). Next to each sketch, <u>record</u> a representative example.

Open system:

Example:____

____.

Closed system:

Example:____.

5. Earth's equatorial diameter is_______ km. Earth's polar diameter is_________ km. What is the reason for this difference?:____

____. In this modern era Earth's shape is described as a____.

6. Properly label the following illustrations of latitude and longitude in the spaces provided.

(a)

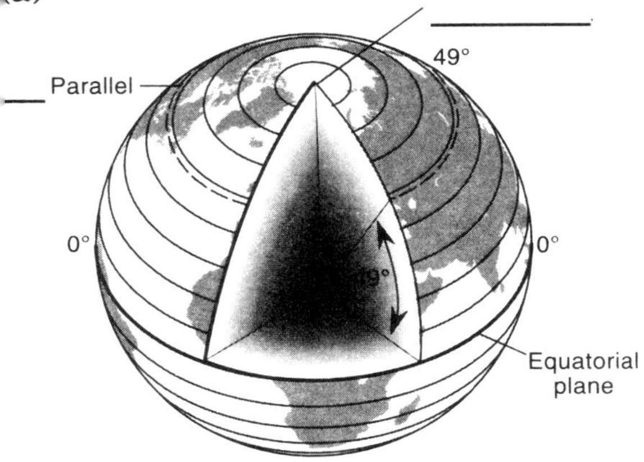

____ Parallel

49°

0°

0°

49°

Equatorial plane

(a) Latitude

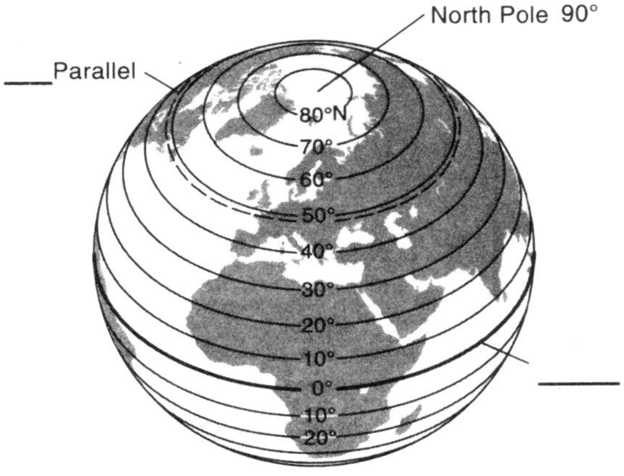

North Pole 90°

____ Parallel

80°N, 70°, 60°, 50°, 40°, 30°, 20°, 10°, 0°, 10°, 20°

(b) Parallels

(b)

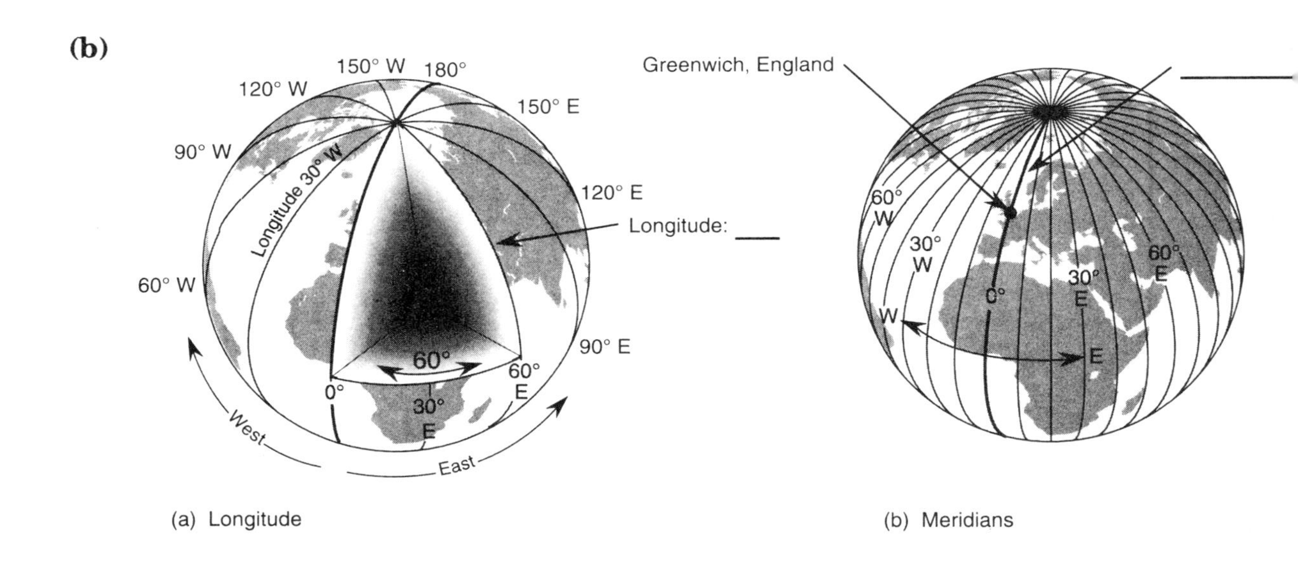

(a) Longitude

(b) Meridians

What is your present latitude:____________________; longitude:____________________.

What was your source for finding these coordinates?____________________________________

____.

7. Relative to Eratosthenes and his determination of Earth's circumference, answer the following questions and complete the labeling of the following figure from FYI Report 1-1 (p. 11-12) on the next page.

(a) How many stadia separated Syene and Alexandria according to Eratosthenes?

____________________. Each stadion represented approximately _____ m.

(b) He determined that this distance was approximately what fraction of Earth's total circumference?__________. Using this information, Earth's circumference is represented by how many stadia?__________; which is equal to how many meters?__________m.

(c) Given Earth's actual equatorial circumference of __________m, what percent error does this represent?__________%.

Eratosthenes' Calculation:

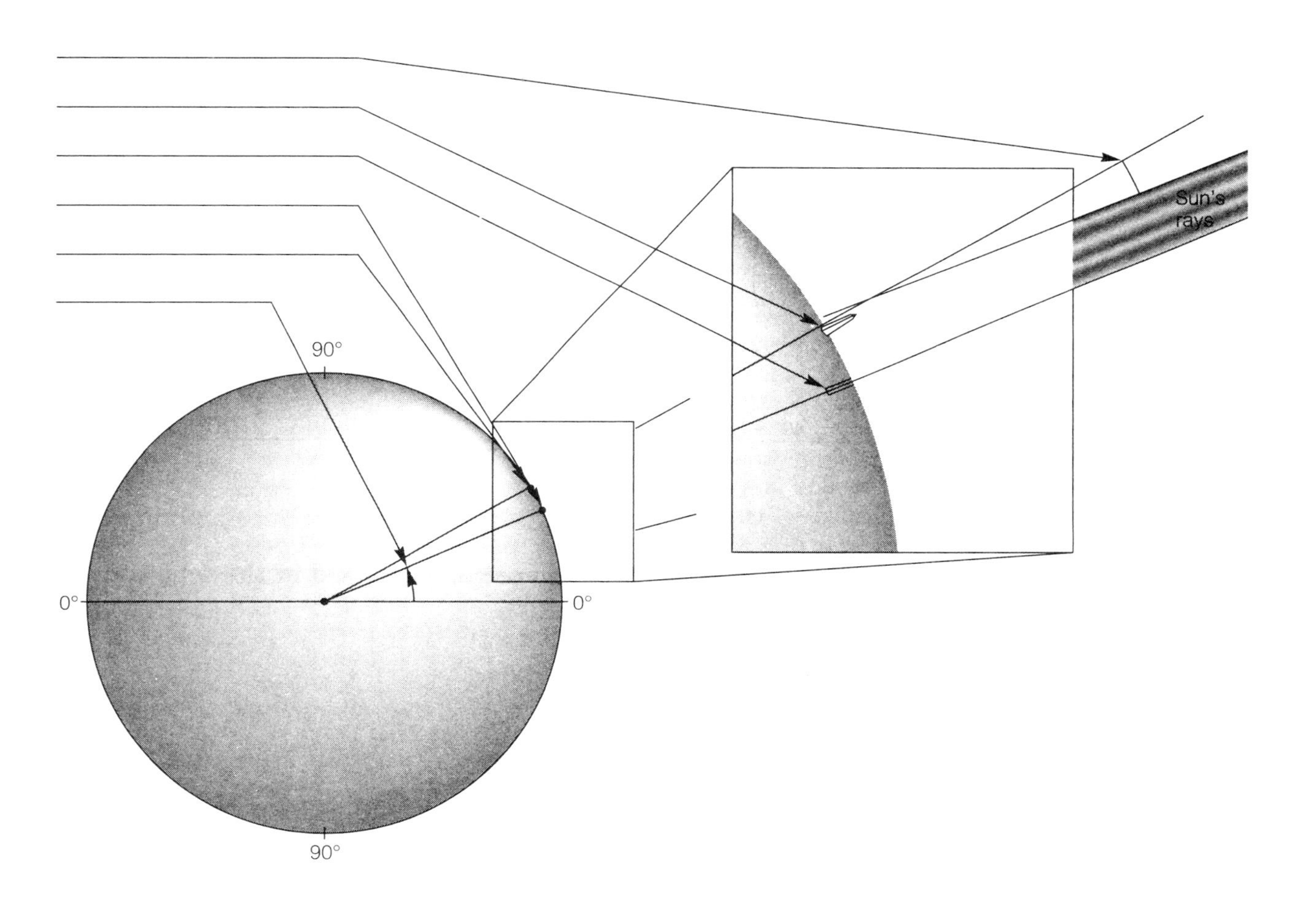

8. According to Figure 1-14 (p. 18) and your understanding of the concepts presented, label or describe all the components in this illustration:

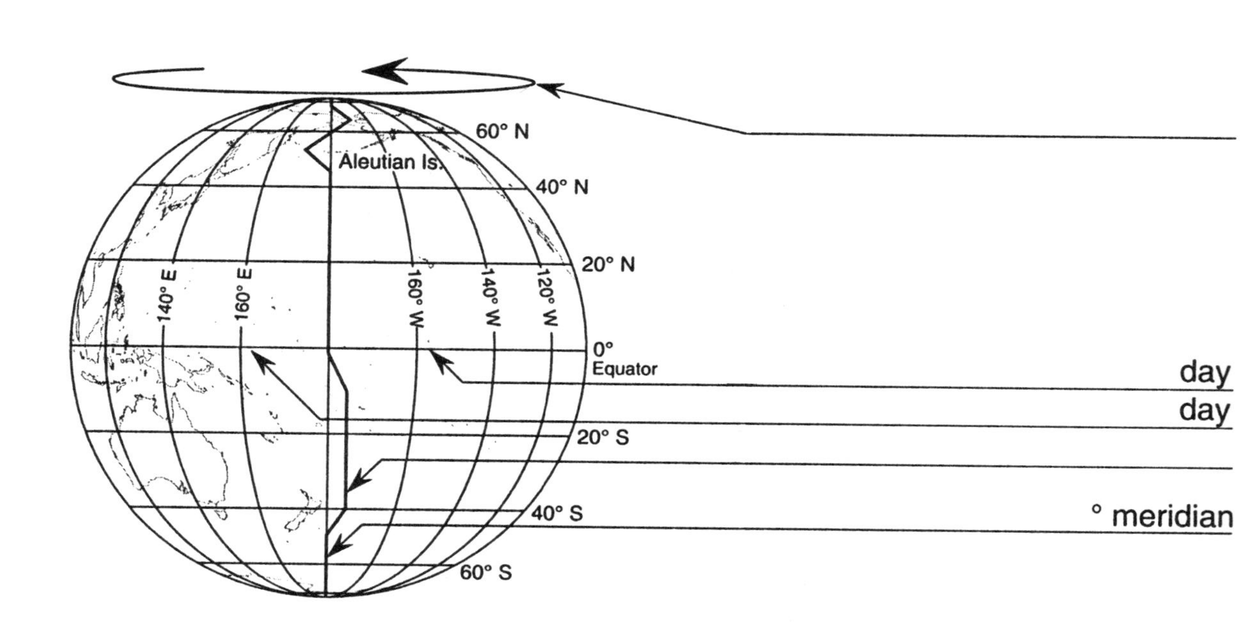

9. Describe and explain each of the following map projections used in the text and their characteristics and properties (pp. 23-25).

(a) Robinson projection: Description:____

____

____.

(b) Goode's homolosine projection: Description:______________________________

____

____.

(c) Mercator projection: Description:____

____

____.

10. Because Earth revolves 360° every 24 hours, or 15° per hour (360° ÷ 24 =15°), a time zone of one hour is established for each 15° of longitude. Each time zone theoretically covers 7.5° on either side of a controlling meridian (0°, 15°, 30°, 45°, 60°, 75°, 90°, 105°, 120°, etc.) and represents one hour.

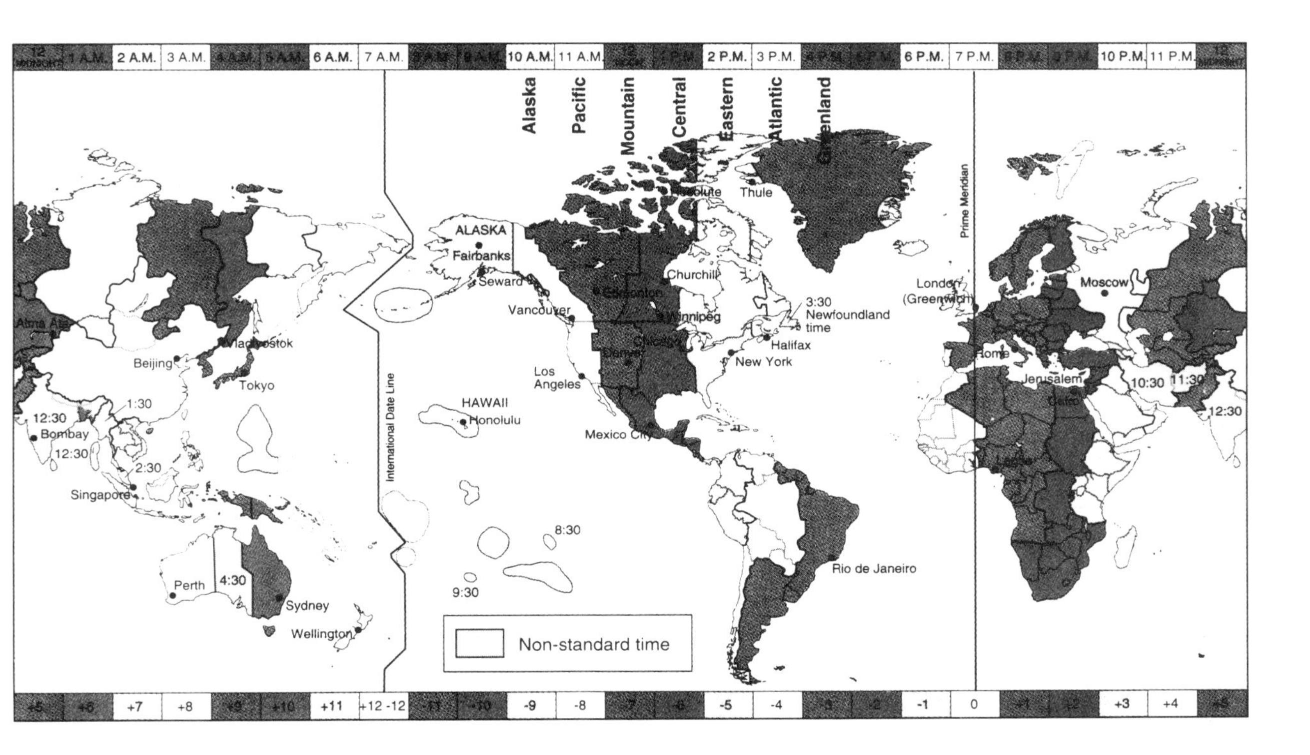

Modern international standard time zones. Numbers indicate how many hours each zone is earlier (minus sign) or later (plus sign) than Greenwich Mean Time, today known as Coordinated Universal Time.

(a) From the map of global time zones above (Figure 1-15, p. 20) can you determine the present time in the following cities? (For your local time use the time at this moment.)

Your present time?__________________	Your time zone:__________________
Moscow? __________________	Denver? __________________
London? __________________	Los Angeles? __________________
Halifax?__________________	Fairbanks?__________________
Chicago? __________________	Honolulu?__________________
Winnipeg?__________________	Singapore?__________________

(b) What is the distance in km (mi) between your college and the standard (controlling) meridian for your time zone?________________________________.

11. What is UTC? Briefly describe.____

____

____.

12. What is GPS? Briefly describe.____

____

____.

13. What is GIS? Briefly describe.____

____

____.

14. Using Figure 1-24, and the description on pp. 27-29, define remote sensing. Distinguish between an active and passive systems and give an example of each.________________

____

____

____.

15. Relative to remote sensing, distinguish a "photograph" from an "image."

(a) Photograph:____

____

____;

(b) Image:____

____.

Sample Self-test

(Answers at the end of the study guide.)

1. The word *spatial* refers to
   a) the how and why questions rather than the where question
   b) items that relate specifically to society
   c) the nature and character of physical space
   d) eras of time, important to history

2. Relative to the fundamental themes of geography authored by the Association of American Geographers, latitude and longitude refer to
   a) location
   b) place
   c) movement
   d) regions

3. The three most costly natural disasters in history in terms of property damage—Hurricane Andrew, the Midwest floods, and the Northridge (Reseda) quake—fall within which theme?
   a) place
   b) human-Earth relationships
   c) movement
   d) regions

4. Earth's environment, operating as a system, is characterized by which one of the following?
   a) an open system in terms of both energy and matter (air, water, soil)
   b) a closed system in terms of energy
   c) entirely abiotic
   d) a closed system in terms of matter (air, water, soil)

5. Earth's circumference was first calculated by
   a) Columbus
   b) modern satellite measurements
   c) Eratosthenes, the librarian at Alexandria
   d) Sir Isaac Newton

6. The basis of time is the fact that Earth
   a) rotates from east to west, or westward
   b) GMT is measured in Washington, D.C.
   c) moves around the Sun taking 365.25 days a year
   d) rotates on its axis in 24 hours, or 15° per hour

7. UTC refers to
   a) Universal Time Conference
   b) Coordinated Universal Time
   c) United States time
   d) a system of local time using phases of the moon

8. Which of the following possesses *all* of Earth's surface properties of area, shape, direction, proximity, and distance
   a) Goode's homolosine projection
   b) Mercator projection
   c) Alber's equal-area conic projection
   d) a globe

9. Geography is defined by its spatial approach rather than by a specific body of knowledge or content.
   a) true
   b) false

10. The tangible and intangible aspects of a region specifically refer to the geographic theme of location.

a) true b) false

11. Negative feedback, most common in nature, tends to discourage response in a system; promoting system self-regulation.

a) true b) false

12. The precise determination of longitude at sea was impossible until as late as A.D. 1760.

a) true b) false

13. That part of geography that embodies mapmaking is called geodesy.

a) true b) false

14. The Eastern Hemisphere side of the International Date Line (180th meridian) is always one day *ahead* of the Western Hemisphere side.

a) true b) false

15. According to the text, only the Goode's homolosine map projection possesses *both* the qualities of equal area and true shape.

a) true b) false

16. Geography is a ________________ science.

17. In terms of energy Earth is an ___________ system, whereas in terms of matter Earth is a ___________ system. A modern landfill (dump) is indicative of the fact that Earth is a ___________ material system.

18. From anywhere in the Northern Hemisphere, you can determine your latitude by sighting on ______________ at night. In the Southern Hemisphere the ________________________ constellation is used to find the celestial pole.

19. Coordinated Universal Time (UTC) is now measured by six___________________________ clocks. The United States clock is named the ____________________, operated by Time and Frequency Services of the N.I.S.T. which stands for:_____________________________.

20. A great circle is____; a small circle is____.

21. The four classes of map projections are:

a) __________________________ c) __________________________

b)__________________________ d) __________________________

PART ONE:
The Energy-Atmosphere System

Overview–Part One

Part One exemplifies the systems organization of the text: Part One of ELEMENTAL GEOSYSTEMS begins with the Sun and solar system. Solar energy passes across space to Earth's atmosphere where it is distributed unevenly as a result of Earth's curvature. Physical factors produce seasonal variations of daylength and Sun altitude as the subsolar point migrates between the tropics. The atmosphere acts as a membrane editing out harmful wavelengths of electromagnetic energy and protecting Earth from the solar wind and most debris from space. The lower atmosphere is being dramatically altered by human activities (Chapter 2). Insolation passes through the atmosphere to Earth's surface where patterns of surface energy budgets are produced. These flow processes produce world temperature patterns (Chapter 3) and general and local atmospheric and oceanic circulations are created (Chapter 4). These three chapters portray the Earth-Atmosphere System.

Name:______________________ Class Section:______________

Date:______________ Score/Grade:__________

Solar Energy, Seasons, and the Atmosphere

2

Chapter Overview

Our planet and our lives are powered by radiant energy from the star closest to Earth—the Sun. In this chapter we follow the Sun's output to Earth and the top of the atmosphere. The uneven distribution of insolation sets the stage for all the motions and flow systems we will study in later chapters.

Think for a moment of the annual pace of your own life, your wardrobe, gardens, and life-style activities—all reflect shifting seasonal energy patterns. Seasonality—the periodic rhythm of warmth and cold, dawn and daylight, twilight and night—affects all of our lives and has fascinated humans for centuries.

Earth's atmosphere is a unique reservoir of gases, the product of billions of years of evolutionary development. This chapter examines the atmosphere's structure, function, and composition. Insolation arrives from outer space and descends through the various layers and regions of the atmosphere. The guiding concept of this chapter involves the flow of insolation from the top of the atmosphere down through the atmosphere to Earth's surface. A consideration of our modern atmosphere must also include the spatial aspects of human-induced gases that affect it, such as air pollution, the stratospheric ozone predicament, and the blight of acid deposition.

Learning Objectives

The following learning objectives help guide your reading and comprehension efforts. The operative word is in *italics*. Read and work with these carefully and note that exercise #1 asks you about five of these objectives. After reading the chapter and using this workbook, you should be able to:

1. *Distinguish* between galaxies, stars/suns, and planets.
2. *Diagram* Earth's orbit about the Sun and *label* the perihelion and aphelion positions.
3. *Explain* the average distance from the Sun using kilometers (miles) and the speed of light.
4. *Describe* the hypotheses that explain the formation of the Sun and planets from the dust, gases, and icy comets of the nebula, i.e., the nebular and planetesimal hypotheses.
5. *Explain* the scientific method and *distinguish* between observation, generalization, discernment of coherent patterns, hypothesis, and eventual theory construction.
6. *Describe* the operation of the Sun and solar fusion.
7. *Distinguish* between the solar wind and the electromagnetic spectrum of radiant energy.
8. *Analyze* the solar wind-magnetosphere interaction and *relate* this to auroral activity in the upper atmosphere.
9. *List* by wavelength the components of the electromagnetic spectrum of radiant energy from the Sun.
10. *Explain* the reason for the uneven distribution of insolation at the thermosphere and the effects this creates in the atmosphere below.
11. *Describe* ancient commemorations of seasons marked by the construction of monuments or calendars.
12. *Define* the Sun's altitude and declination and *describe* the annual variability of each and their role in seasonality.
13. *Explain* the annual variability of daylength and its role in seasonality.
14. *List* the five contributing factors that are reasons for seasons.
15. *Construct* a simple sketch of Earth's revolution about the Sun and explain what is meant by the "annual march of the seasons."
16. *Describe* Earth's rotation on its axis and *differentiate* the speed of that rotation by latitude.
17. *Portray* the Sun's declination, seasonal anniversary dates, and the equation of time using the analemma.
18. *Describe* Earth's tilt in relation to the plane of the ecliptic.
19. *Contrast* seasonal conditions at each of the four seasonal anniversary dates during the year and *list* the name given and the Sun's declination for each of these dates.
20. *Contrast* a cell membrane with Earth's atmosphere as described in the text and as used in the quote from Lewis Thomas' *The Lives of a Cell*.
21. *Construct* a general model of the atmosphere based on composition: heterosphere and homosphere.
22. *Define* layers within this model based on temperature: thermosphere, mesosphere, stratosphere, and troposphere; and function: ionosphere and ozonosphere (ozone layer).
23. *Diagram* this atmospheric model in an integrated fashion similar to Figure 2-19 (p. 55).

24. *List* the stable components of the modern atmosphere and their relative percentage contributions by volume.
25. *Describe* conditions within the stratosphere: composition, temperatures, and function.
26. *Analyze* the destructive nature of chlorofluorocarbons (CFCs) on the ozonosphere and *describe* the reactions determined by the work of scientists Rowland and Molina and present scientific measurements.
27. *Explain* the political actions and inactions concerning the present condition of the ozone layer.
28. *Distinguish* between natural and anthropogenic variable gases and materials in the lower atmosphere.
29. *Contrast* a temperature inversion with a normal temperature lapse-rate profile in the lower troposphere.
30. *List* the anthropogenic gases in the lower atmosphere and *identify* the symbol and the principal source for each.
31. *Describe* the health effects of CO, NO_2, O_3, and SO_2 on humans.
32. *Construct* a simple equation that illustrates photochemical reactions and the production of NO_2, O_3, and PAN.
33. *Portray* the role of fossil fuels and transportation in the production of air pollution, and *relate* this to the formation of acid deposition compounds.

Outline Headings and Glossary Review

These are the first- second-, and third-order headings that divide this chapter. The key terms and concepts that appear **boldface** in the text are listed under their appropriate heading in bold italics; these highlighted terms appear in the text glossary. A check-off box is placed next to each key term so you can mark your progress through the chapter as you define these in your reading notes or prepare note cards.

The Solar System, Sun, and Earth

- ❑ ***Milky Way Galaxy***
- ❑ ***nebula***
- ❑ ***planetesimal hypothesis***
- ❑ ***scientific method***

Dimensions and Distances

- ❑ ***speed of light***

Earth's Orbit

- ❑ ***perihelion***
- ❑ ***aphelion***

Solar Energy: From Sun to Earth

- ❑ ***fusion***

Solar Wind

- ❑ ***solar wind***
- ❑ ***sunspots***
- ❑ ***magnetosphere***
- ❑ ***auroras***

Electromagnetic Spectrum of Radiant Energy

- ❑ ***electromagnetic spectrum***
- ❑ ***wavelength***

Energy at the Top of the Atmosphere

- ❑ ***thermopause***

Intercepted Energy

- ❑ ***insolation***

Solar Constant

- ❑ ***solar constant***

Uneven Distribution of Insolation

- ❑ ***subsolar point***

Global Net Radiation

The Seasons

Seasonality

- ❑ ***altitude***
- ❑ ***declination***
- ❑ ***daylength***

Reasons for Seasons

Revolution

- ❑ ***revolution***

Rotation

- ❑ ***rotation***
- ❑ ***axis***
- ❑ ***circle of illumination***

Tilt of Earth's Axis

- ❑ ***plane of the ecliptic***

Axial Parallelism

- ❑ ***axial parallelism***

Annual March of the Seasons

- ❑ ***winter solstice***
- ❑ ***December solstice***
- ❑ ***Tropic of Capricorn***

- ❑ ***vernal equinox***
- ❑ ***March equinox***
- ❑ ***summer solstice***
- ❑ ***June solstice***
- ❑ ***Tropic of Cancer***
- ❑ ***autumnal equinox***
- ❑ ***September equinox***

Seasonal Observations

Dawn and Twilight Concepts

Earth's Atmosphere

Heterosphere

- ❑ ***heterosphere***
- ❑ ***exosphere***

Thermosphere

- ❑ ***thermosphere***
- ❑ ***kinetic energy***
- ❑ ***sensible heat***

Ionosphere

- ❑ ***ionosphere***

Homosphere

- ❑ ***homosphere***
- ❑ ***carbon dioxide***

Mesosphere

- ❑ ***mesophere***

Stratosphere and Ozonosphere

- ❑ ***stratosphere***
- ❑ ***ozonosphere***
- ❑ ***ozone layer***

Troposphere

- ❑ ***troposphere***
- ❑ ***normal lapse rate***
- ❑ ***environmental lapse rate***

Variable Atmospheric Components

Natural Sources

Natural Factors that Affect Air Pollution

Temperature Inversion

- ❑ ***temperature inversion***

Anthropogenic Pollution

Carbon Monoxide

- ❑ ***carbon monoxide***

Photochemical Smog Reactions

- ❑ ***photochemical smog***
- ❑ ***nitrogen dioxide***
- ❑ ***peroxyacetyl nitrates (PAN)***

Ozone Pollution

Industrial Smog and Sulfur Oxides

- ❑ ***industrial smog***
- ❑ ***sulfur dioxide***

SUMMARY

Learning Activities and Critical Thinking

1. Select any five learning objectives from the list presented at the beginning of this chapter. Place the number selected in the space provided (no need to rewrite each objective). Using the following questions as guidelines only, briefly discuss your treatment of the objective.

- What did you know about the objective before you began?
- What was your plan to complete the objective?
- Which information source did you use in your learning (text, or other)?
- Were you able to complete the action stated in the objective? What did you learn?
- Are there any aspects of the objective about which you want to know more?

a)____:____

____

____

____

____

____.

b)_____:____
____
____
____
____
____.

c)_____:____
____
____
____
____
____.

d)_____:____
____
____
____
____
____.

e)_____:____
____
____
____
____
____.

2. Referring to Figure 2-1 and your knowledge, rank the nine planets in the following categories: **a)** distance from the Sun (closest to farthest), and, **b)** diameter (from largest to smallest estimated from illustration or an encyclopedia).

a)	**b)**
1.____________________	1.____________________
2.____________________	2.____________________
3.____________________	3.____________________
4.____________________	4.____________________
5.____________________	5.____________________
6.____________________	6.____________________
7.____________________	7.____________________
8.____________________	8.____________________
9.____________________	9.____________________

3. In the space provided, make a simple sketch of Earth's orbit about the Sun; identify *perihelion* and *aphelion* locations, dates, and orbital distances (Figure 2-2, p. 40).

S

4. A portion of the electromagnetic spectrum of radiant energy is emitted by the Sun as shown in Figure 2-7 (p. 43). On the following illustration, complete the labeling by identifying wavelengths, naming the portions of the spectrum, and identifying the components of visible light.

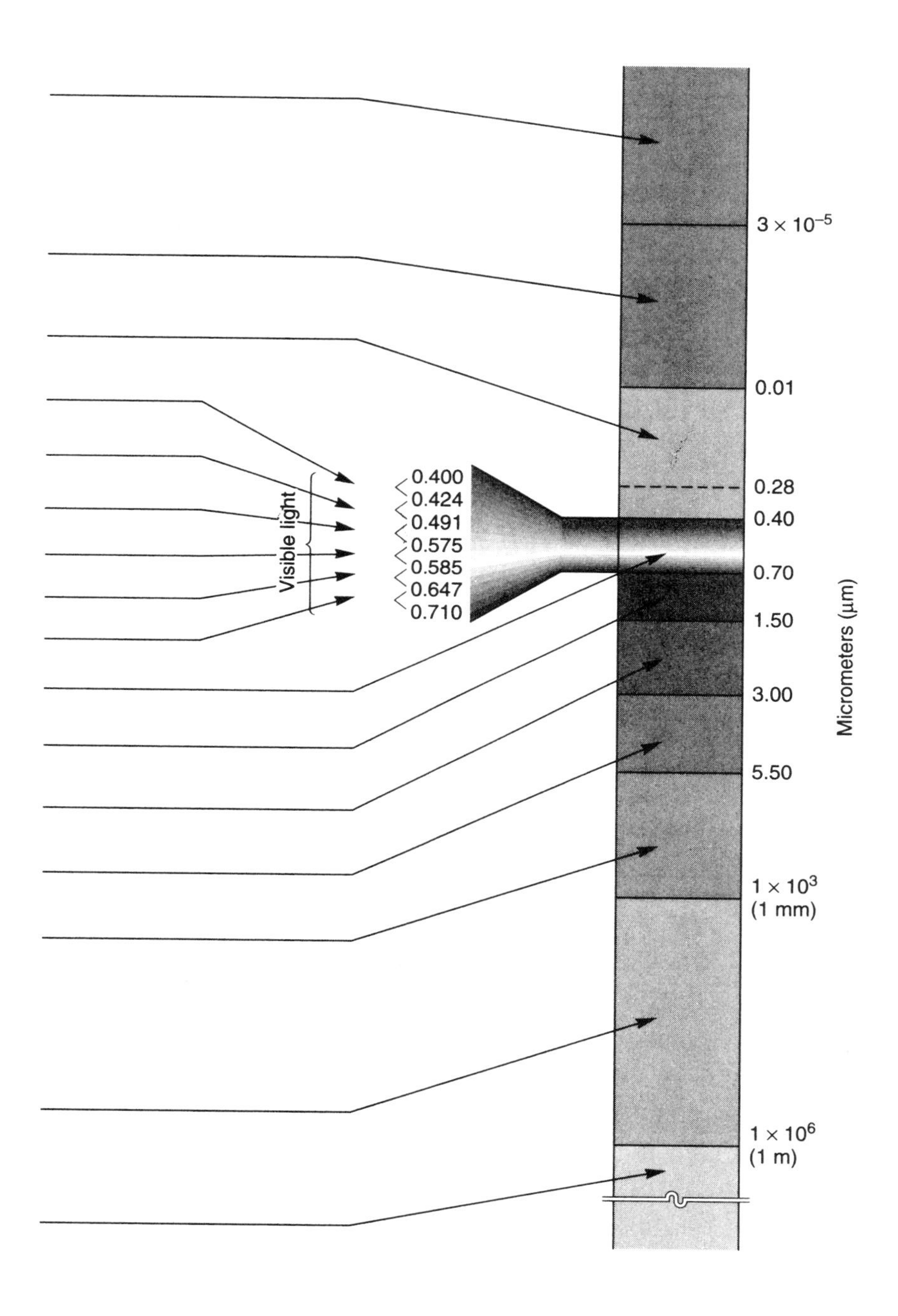

5. Describe in your own words the intercepted energy at the top of the atmosphere. Include discussion of the *thermopause*, *insolation*, *solar constant*, and *subsolar point* on the surface below.

____

____

____

____

____.

6. Why is this energy input into the Earth-atmosphere system unevenly received by latitude? Explain Figure 2-9, p. 46, in your own words.

____

____

____

____.

7. Figure 2-10, p. 47, presents average daily net radiation flows at the top of the atmosphere. Figure 2-11, p. 48, summarizes energy surpluses and deficits for the entire Earth-atmosphere system. How do these two figures tie together? What do they tell you about Earth's energy budget? What can you infer about the patterns of climate and vegetation at the surface produced by these energy receipts.

____

____

____

____

____.

8. Summarize the following seasonal anniversary dates.

Approx. date	**Name**	**Subsolar point location (declination)**	**Daylength at North Pole**
December			
March			
June			
September			

9. Table 2-1 (p. 49) lists the five interacting reasons for seasons. In the space provided, describe and explain why each affects seasonality.

Factor	Specific effect on seasonality
a)__________	____________________ .
b)__________	____________________ .
c)__________	____________________ .
d)__________	____________________ .
e)__________	____________________ .

10. Describe Earth's rotation as completely as possible.____________________

____

____.

11. The text classifies the atmosphere according to three criteria. Identify the regions of the atmosphere defined by each criteria and record their physical extent in km (or mi) in the spaces provided. (Consult pp. 54-58 and Figure 2-19.)

Criteria	Atmospheric Region	Extent in km (mi)
Composition:	**a)**	
	b)	
Temperature:	**a)**	
	b)	
	c)	
	d)	
Function:	**a)**	
	b)	

12. Figure 2-19, p. 55, and Figure 4-6, p. 112, when placed together form an integrated illustration that identifies the criteria for classification of the atmosphere, temperature and pressure profiles with altitude, as well as other information. Use the reproduction of this figure below to <u>record</u> all the elements and complete the labels as these figures appear in the text. The best way to work with this is for you to add additional information from the text, the labels, and coloration (color pencils are best) as you read the chapter.

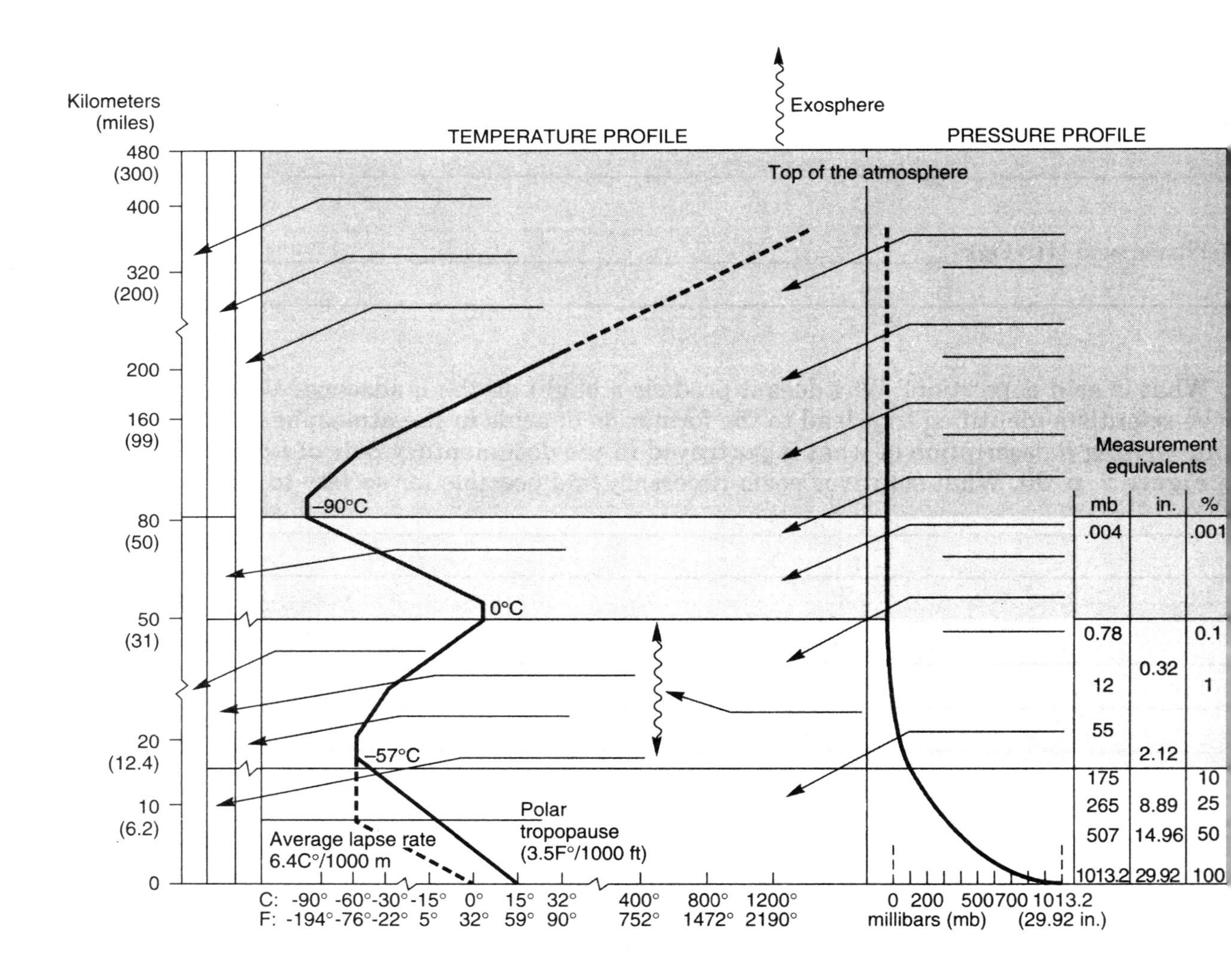

13. **(a)** Describe in your own words the predicament relative to stratospheric ozone, the identification and history of the situation, and its causes, current action being taken, and present status. **(b)** What is your interpretation of Figure 2, p. 61, in the text. Explain. **(c)** What does this mean to you personally? Do you have any action or response plans related to this situation?

(a)______________________________;

(b)______________________________;

(c)______________________________;

What is the "UV Index" (p. 58)?____________________.

14. Using Figure 2-26, p. 67, and the text (pp. 65-68), describe how each of the following photochemical pollutants is formed in the lower troposphere. Simple generalizations of the formula expressions in the illustration are all right. Assume the source of NO_2 from transportation systems and the presence of ultraviolet light in sunlight.

a) Ozone (O_3): ____

____

____.

b) PAN (peroxyacetyl nitrates): ____

____

____.

c) Nitric acid (HNO_3): ____

____

____.

Sample Self-test

(Answers appear at the end of the study guide.)

1. Our planet and our lives are powered by
   a) energy derived from Earth's systems
   b) radiant energy from the star closest to Earth
   c) utilities and oil companies
   d) the solar wind

2. In terms of its distance from the Sun Earth is
   a) the farthest planet of the nine
   b) the closest planet of the nine
   c) about 50,000 km (30,000 mi) away
   d) closer in January and farther in July

3. The Sun produces which of the following?
   a) visible light only
   b) streams of charged particles and radiant energy
   c) only the solar wind
   d) only radiant energy

4. The auroras in the upper atmosphere are caused by
   a) the interaction of electromagnetic radiant energy with atmospheric gases
   b) Earth-generated radio broadcasts
   c) various weather phenomena
   d) the interaction of the solar wind and atmospheric gases

5. Intercepted solar radiation is called
   a) solar wind
   b) light
   c) thermosphere
   d) insolation

6. The scientific method is described by which of the following?
   a) the application of common sense
   b) the development of hypotheses for testing and prediction
   c) the formation of theories encompassing broad general principles
   d) all of these are correct

7. Which of the following is false?
   a) maps are not important to the development of a GIS
   b) GIS stands for geographic information system
   c) satellite weather images are an example of remote sensing
   d) GIS represents an important planning tool for geographers and others

8. Changes in *daylength* and the Sun's *altitude* represent
   a) revolution
   b) phenomena that occur only at the equator
   c) the concept of seasonality
   d) factors that remain constant and unchanging throughout the year

9. The Sun's declination refers to
   a) the angular distance from the equator to the point where direct overhead insolation is received (the subsolar point)
   b) the angular difference between the horizon and the Sun
   c) the Sun's tilt on its axis
   d) its altitude in the sky

10. The Tropic of Capricorn refers to
    a) the parallel that is 23.5° south latitude
    b) the location of the subsolar point on September 22
    c) the parallel at the farthest northern location for the subsolar point during the year
    d) the location of the subsolar point during the vernal equinox

11. On June 21st, the Sun's declination is at
    a) the equator
    b) Rio de Janeiro, Brazil and Alice Springs, Australia
    c) the Tropic of Capricorn
    d) the Tropic of Cancer

12. Lewis Thomas, in his book, *The Lives of the Cell*, compared Earth's atmosphere with a/an
    a) layer of human skin
    b) membrane of a cell
    c) layer of water
    d) covering of soil over a rock

13. Based on <u>composition</u>, the atmosphere is divided into
    a) five regions beginning with the outermost thermosphere
    b) two broad regions
    c) nitrogen and oxygen
    d) the troposphere and the stratosphere

14. The region of the atmosphere that is so evenly mixed that it behaves as if it were a single gas is
    a) the homosphere
    b) the heterosphere
    c) the exosphere
    d) the thermosphere

15. Which of the following lists of gases is correct, from <u>most to least</u> in terms of percentage within the homosphere?
    a) nitrogen, argon, oxygen, xenon, carbon dioxide
    b) nitrogen, oxygen, argon, carbon dioxide, trace gases
    c) oxygen, PAN, ozone, nitrogen, carbon dioxide
    d) water vapor, oxygen, argon, carbon dioxide

16. Recent measurements of increased levels of ultraviolet light at Earth's surface
    a) are focused along the equator and not the polar regions
    b) are related to an increasing rate of skin cancer that is rising at 10% per year
    c) are unrelated to stratospheric ozone
    d) affect those at sea level more than those living in mountains

17. Relative to lapse rates in the troposphere,
    a) the <u>environmental</u> lapse rate refers to the actual lapse rate at any particular time and may vary greatly from the normal lapse rate
    b) temperatures tend to increase with altitude
    c) temperatures remain constant with increasing altitude
    d) an average normal lapse rate value of 10C° per 1000 m of altitude increase is used

18. PAN in the lower troposphere
    a) is principally related to sulfur dioxides
    d) is formed by particulates such as dust, dirt, soot, and ash
    c) damages and kills plant tissue, a photochemical product
    d) comes directly from automobile exhaust

19. Industrial smog is
    a) associated with photochemistry
    b) principally associated with coal-burning industries
    c) a relatively recent problem during the latter half of this century
    d) principally associated with transportation

20. The oxides of sulfur and nitrogen
    a) lead to the formation of airborne sulfuric and nitric acid
    b) form acids that are deposited in both dry and wet forms
    c) are produced by industry and transportation
    d) all of these choices are true

21. The Solar System is located on a remote trailing edge of the Milky Way Galaxy.
    a) true
    b) false

22. It takes sunlight 24 hours to reach the top of the atmosphere from the Sun at the speed of light.
    a) true
    b) false

23. Earth is closest to the Sun in January and farther away in July.
    a) true
    b) false

24. Suns condense out of nebular clouds of dust and gas, and planets form by the accretion of dust, gas, and icy comets in the solar nebula.
    a) true
    b) false

25. The average amount of energy received at the thermopause is 1372 watts per m^2 (2 $cal/cm^2/min$).
    a) true
    b) false

26. The Sun emits radiant energy that is composed of 8% ____ wavelengths; 47% ____________________ wavelengths; and 45% ____________________ wavelengths.

27. GIS stands for____.

28. The two layers of the atmosphere specifically defined according to their <u>function</u> are the troposphere and the heterosphere.
    a) true
    b) false

29. In the lower atmosphere, nitrogen is essentially a by-product of photosynthesis, whereas oxygen is a product of bacterial action.
    a) true   b) false

30. The ozonosphere is presently under attack by photochemical smog.
    a) true   b) false

Name:______________________________ Class Section:________________

Date:________________ Score/Grade:______________

Atmospheric Energy and Global Temperatures

Chapter Overview

Earth's biosphere pulses with flows of energy. Chapter 3 follows the passage of solar energy through the lower atmosphere to Earth's surface. This entire process is a vast flow-system with energy cascading through fluid Earth systems. An integrated illustration, Figure 3-6, presents the Earth-atmosphere energy balance; referencing this figure as you read the chapter is helpful. The chapter then analyzes surface energy budgets and develops the concept of net radiation.

Air temperature has a remarkable influence upon our lives, both at the microlevel and at the macrolevel. A variety of temperature regimes worldwide affect entire lifestyles, cultures, decision-making, and resources spent. Global temperature patterns appear to be changing in a warming trend that is affecting us all and is the subject of much scientific, geographic, and political interest. Our bodies sense temperature and subjectively judge comfort, reacting to changing temperatures with predictable responses. These are some of the topics addressed in Chapter 3.

This chapter presents concepts that are synthesized on three temperature maps: January, July, and the annual range. As the chapter develops, reference to these maps will be useful. The chapter relates these temperature concepts directly to you with a discussion of apparent temperatures—the wind chill and heat index charts. In addition, the energy characteristics of urban areas are explored, for the climates in our cities differ measurably from those of surrounding rural areas.

Learning Objectives

The following learning objectives help guide your reading and comprehension efforts. The operative word is in *italics*. Read and work with these carefully and note that exercise #1 asks you about five of these objectives. After reading the chapter and using this workbook, you should be able to:

1. *Identify* the pathways of energy transmission through the troposphere to Earth's surface: transmission, scattering (diffuse reflection), diffuse radiation, refraction, convection, advection, and conduction.
2. *Compare* the albedo values of different surfaces and *utilize* these values to determine the differential heating of these surfaces.
3. *Analyze* the pattern of average annual solar radiation receipt on a horizontal surface at ground level.
4. *Discern* the difference between an actual greenhouse and the greenhouse effect in the troposphere.

5. *Explain* the various ways human activities alter aspects of the Earth-atmosphere energy balance.
6. *Review* the pathways of terrestrial infrared radiation from Earth's surface, through the atmosphere, and eventually to space—Earth's net outgoing longwave.
7. *Diagram* the daily radiation curves for Earth's surface and *label* the key aspects of incoming, outgoing, and net infrared radiation and the daily temperature lag.
8. *Construct* a simple equation to demonstrate the derivation of each component that produces surface net radiation (NET R).
9. *Compare* and *contrast* the energy balance for El Mirage, California, and Pitt Meadows, British Columbia.
10. *Review* the technologies, capabilities, and working examples of solar-electric and solar-thermal energy resources.
11. *Define* the concept of temperature.
12. *Distinguish* between Kelvin, Celsius, and Fahrenheit scales and *describe* their origins.
13. *Explain* the different ways temperature is measured and the operation of a minimum-maximum thermometer.
14. *Describe* a standard weather instrument shelter.
15. *List* the principal controls and influences that produce global temperature patterns.
16. *Review* the role of latitudinal location and the amount of insolation received.
17. *Explain* the concept of normal lapse rate and *diagram* a simple drawing illustrating the concept.
18. *Portray* the behavior of air temperature at higher elevations for day/night and sunlight/shadow and *explain* why this contrast occurs.
19. *Describe* the elevation of snowlines in mountains at different latitudes.
20. *Contrast* and *compare* the temperature patterns of La Paz (high elevation) and Concepción (low elevation), Bolivia.
21. *Explain* the role of cloud cover and relate its effect on temperature.
22. *Differentiate* between the effects of evaporation over land and water surfaces and its effect on air temperature.
23. *Describe* evaporative cooling; *explain* transmissibility through water as compared to land surfaces.
24. *Define* specific heat and *explain* why this affects land and water heating differences.
25. *Relate* the role of ocean currents to temperature patterns.
26. *Review* the concepts of marine vs. continentality as they influence temperature patterns; *utilize* several pairs of stations to illustrate these differences.
27. *Interpret* the pattern of Earth's temperatures from their portrayal on January and July temperature maps.
28. *Define* apparent temperature and *relate* specific physiological effects of low temperatures and high temperatures on the human body.
29. *Utilize* the wind-chill and heat-index charts to determine some apparent temperatures.
30. *Portray* typical urban heat island conditions.
31. *Contrast* urban areas with surrounding rural environments using several climatic elements.
32. *Diagram* insolation, wind movements, and radiation patterns within an urban dust dome.

Outline Headings and Glossary Review

These are the first- second-, and third-order headings that divide this chapter. The key terms and concepts that appear **boldface** in the text are listed under their appropriate heading in bold italics; these highlighted terms appear in the text glossary. A check-off box is placed next to each key term so you can mark your progress through the chapter as you define these in your reading notes or prepared note cards.

Energy Balance in the Troposphere

Energy in the Atmosphere: Some Basics

Insolation Input

Albedo and Reflection

- ❑ ***reflection***
- ❑ ***albedo***

Scattering (Diffuse Reflection)

- ❑ ***scattering***
- ❑ ***diffuse radiation***

Refraction

- ❑ ***refraction***

Absorption

- ❑ ***absorption***

Earth Reradiation and the Greenhouse Effect

- ❑ ***greenhouse effect***
- ❑ ***convection***
- ❑ ***advection***
- ❑ ***conduction***

Earth-Atmosphere Radiation Balance

Energy at the Surface

Daily Radiation Curves

Simplified Surface Energy Balance

- ❑ ***microclimatology***
- ❑ ***net radiation***

Net Radiation

Principal Temperature Controls

- ❑ ***temperature***

Latitude

Altitude

Cloud Cover

Land-Water Heating Differences

- ❑ ***land-water heating differences***

Evaporation

Transparency

- ❑ ***transparency***

Specific Heat

- ❑ ***specific heat***

Movement

Ocean Currents and Sea-Surface Temperatures

- ❑ ***Gulf Stream***

Summary of Marine vs. Continental Conditions

- ❑ ***marine***
- ❑ ***continentality***

Earth's Temperature Patterns

- ❑ ***isotherm***

January Temperature Map

- ❑ ***thermal equator***

July Temperature Map

Annual Range of Temperatures

Air Temperature and the Human Body

- ❑ ***apparent temperature***
- ❑ ***wind chill factor***

The Urban Environment

- ❑ ***urban heat island***
- ❑ ***dust dome***

SUMMARY

Learning Activities and Critical Thinking

1. Select any five learning objectives from the list presented at the beginning of this chapter. Place the number selected in the space provided (no need to rewrite each objective). Using the following questions as guidelines only, briefly discuss your treatment of the objective.

- What did you know about the objective before you began?
- What was your plan to complete the objective?
- Which information source did you use in your learning (text, or other)?
- Were you able to complete the action stated in the objective? What did you learn?
- Are there any aspects of the objective about which you want to know more?

a)____:____________________

b)____:____________________

c)____:____________________

d)____:____________________

e)____:____________________

2. In the visible wavelengths, ____________colors have lower albedos, and ____________colors have higher albedos. On water surfaces, the angle of the solar rays also affects albedo values; ____________angles produce a greater reflection than do ____________angles. In addition, ____________ surfaces increase albedo, whereas ____________________surfaces reduce it. Record albedos percentage reflected in the spaces below.

Fresh snow:	____________
Forests;	____________
Crops, grasslands:	____________
Concrete:	____________
Asphalt, black top:	____________
Moon surface:	____________
Earth (average):	____________

3. Figure 3-6, p. 101, illustrates the Earth-atmospheric energy balance. As you read through the text, <u>fill</u> <u>in</u> the labels for the pathways of energy transmission. The units of energy involved are added on this version. Please use color pencils to shade in areas with appropriate coloration.

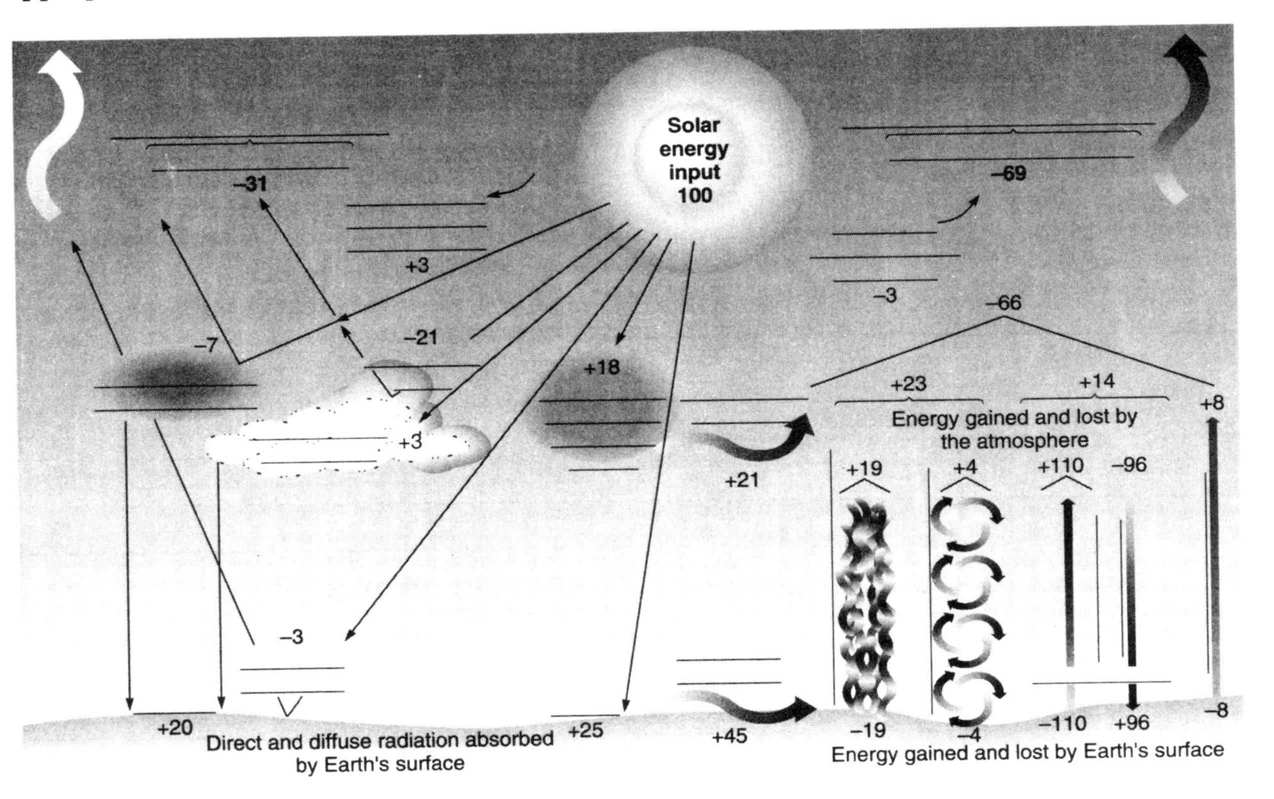

4. In the space provided record a simplified surface energy budget (p. 82), and identify each of the components.

____

5. Net radiation (NET R) is expended in three general output paths. List these expenditures and briefly explain each.

a)____

____;

b)____

____;

c)____

____.

6. Relative to FYI Report 3-1 "Solar Energy Collection and Concentration" (pp. 84-5), describe the nature of the solar energy resource and at least two methods of collection and conversion. In your opinion what appears to be the major stumbling block toward implementation?

____

____

____

____

____

____.

7. In Figure 3-9a and b two photographic examples of different energy environments are presented. Specifically, what do these scenes demonstrate about the climates of the two locations? Which energy component do you think varies greatest between the two stations?

a) El Mirage, CA, as a desert, subtropical location:________________

____

____

____.

b) Pitt Meadows, B.C., as a vegetated, moist, midlatitude location:________________

____

____

____.

8. Using the table inside the back cover of the text, complete the following conversions. Note the position of the degree symbol and the related explanation in Figure 1, p. 87.

25°C________	31°F________	29C°________	–41F°________
4°C________	89°F________	61C°________	53F°________
39°C________	46°F________	26C°________	5F°________
–40°C ________	14°F________	–5C°________	12F°________

9. What temperature is it as you work on this chapter of the workbook?

(a) outdoor temperature? _______°C _______°F

(b) indoor temperature? _______°C _______°F

10. List the four physical aspects of land and water that produce their different responses to heating and therefore different temperatures (Figure 3-11, pp. 90-3).

Water (marine) factors	Land (continental) factors
a)	e)
b)	f)
c)	g)
d)	h)

11. Contrast and compare the marine temperature regime of San Francisco with the continental regime of Wichita. What significant differences do you note?

____

____

____

____

____.

12. Locate Portland, Oregon, and Chicago, Illinois, on the January and July average temperature maps (Figures 3-15 and 3-17, p. 94, 96) and on the annual range of temperature map (Figure 3-18, p. 97). Identify these temperatures for each of the cities (in °C and °F) through *interpolation* using the isotherms on the map.

Portland, Oregon: Jan____________ July_________ Annual Range________________.

Chicago, Illinois: Jan____________ July_________ Annual range________________.

13. Using the graphs on the next two pages, plot the temperature (line graph) and precipitation (bar graph) data for both Portland, Oregon, and Chicago, Illinois. Label each temperature plot and precipitation plot and fill in the city information.

A *climograph* is a graph on which temperature, precipitation, and other weather information is plotted. The following is an example of how your final climographs should appear.

Climograph example:

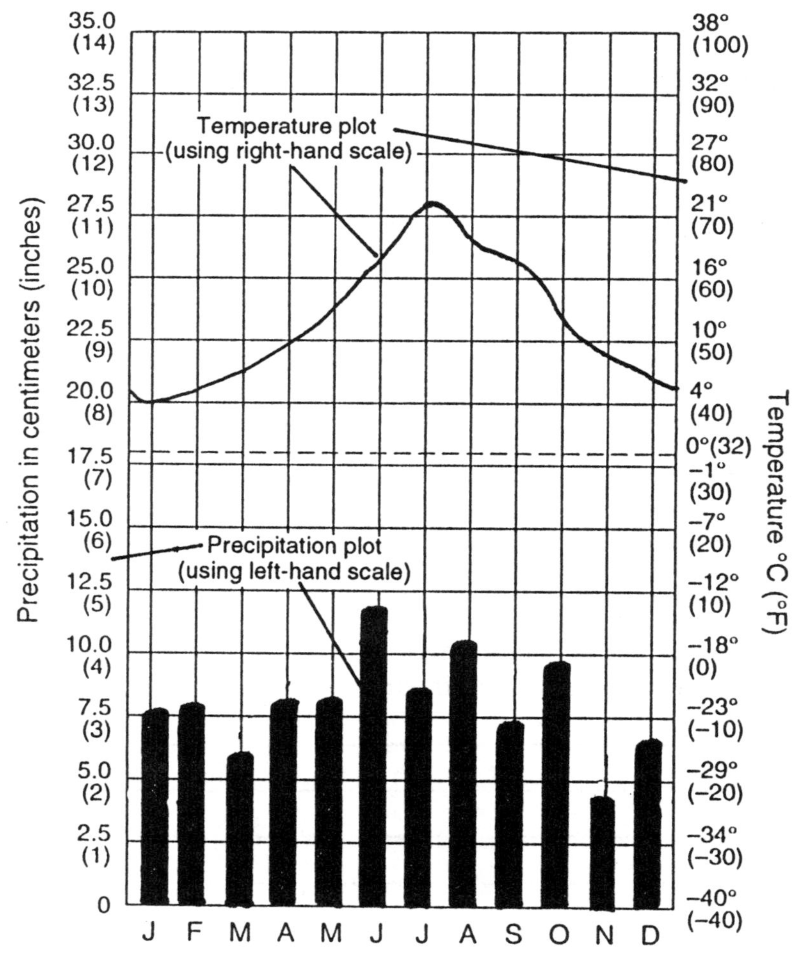

Portland, Oregon

Mediterranean dry, cool summer (Köppen climate symbol: Csb)

Latitude____________________

Longitude____________________

Elevation____________________

Population____________________

Total annual rainfall:____________________

Average annual temperature:____________________

Annual temperature range:____________________

Distribution of temperature during the year:

____

____.

Distribution of precipitation during the year:

____

____.

Precipitation in centimeters (inches): 35.0 (14), 32.5 (13), 30.0 (12), 27.5 (11), 25.0 (10), 22.5 (9), 20.0 (8), 17.5 (7), 15.0 (6), 12.5 (5), 10.0 (4), 7.5 (3), 5.0 (2), 2.5 (1), 0

Temperature °C (°F): 38° (100), 32° (90), 27° (80), 21° (70), 16° (60), 10° (50), 4° (40), 0°(32), −1° (30), −7° (20), −12° (10), −18° (0), −23° (−10), −29° (−20), −34° (−30), −40° (−40)

J F M A M J J A S O N D

Months

Mediterranean Dry, Cool Summer (Csb)

Portland, Oregon: pop. 366,000, lat. 45° 31'N, long. 122 ° 40'W, elev. 9 m (30 ft).

	Jan	Feb	Mar	Apr	May	Jun	Jul	Aug	Sep	Oct	Nov	Dec	Annual
Temperature °C	4.4	6.7	8.9	12.2	15.0	17.8	20.0	20.0	17.8	13.3	8.3	5.6	12.8
(°F)	(40.0)	(44.0)	(48.0)	(54.0)	(59.0)	(64.0)	(68.0)	(68.0)	(64.0)	(56.0)	(47.0)	(42.0)	(55.0)
PRECIP cm	13.7	12.4	10.7	6.1	4.8	4.1	1.0	1.5	4.6	8.9	15.2	18.0	101.3
(in.)	(5.4)	(4.9)	(4.2)	(2.4)	(1.9)	(1.6)	(0.4)	(0.6)	(1.8)	(3.5)	(6.0)	(7.1)	(39.9)

Köppen climate classification symbol:________; name:______________________________

__________________; explanation for this determination:______________________________

____

Chicago, Illinois

Humid continental, hot summer (Köppen climate symbol Dfa)

Latitude________________________________

Longitude________________________________

Elevation________________________________

Population________________________________

Total annual rainfall:________________________

Average annual temperature:__________________

Annual temperature range:____________________

Distribution of temperature during the year:

____

____.

Distribution of precipitation during the year:

____

____.

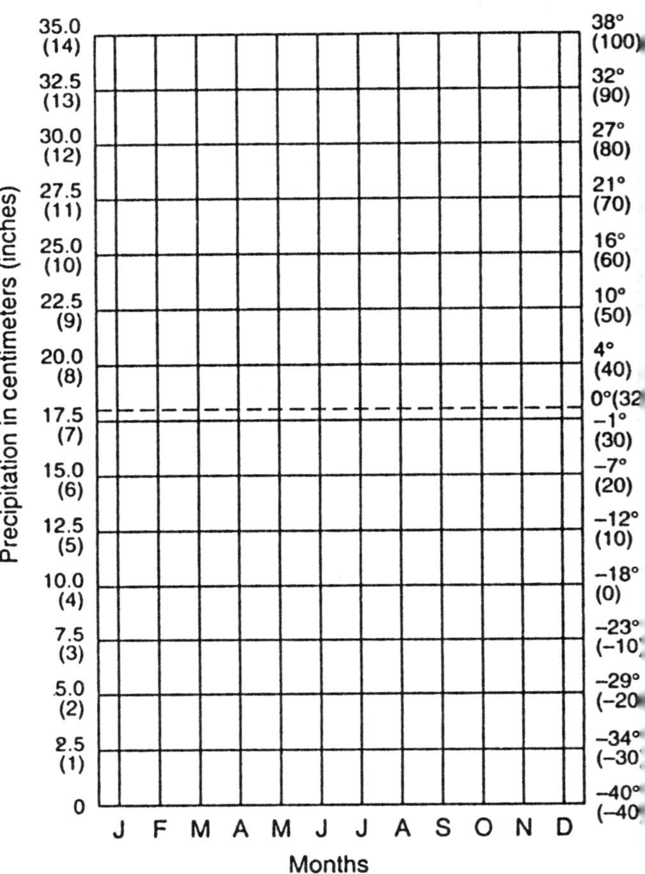

Humid continental, hot summer (Dfa)

Chicago, Illinois: pop. 3,005,000, lat. 41° 47'N, long. 87° 45'W, elev. 186 m (610 ft).

	Jan	Feb	Mar	Apr	May	Jun	Jul	Aug	Sep	Oct	Nov	Dec	A
Temperature°C	-3.9	-2.8	2.3	8.6	14.4	20.0	23.2	22.4	18.7	12.4	4.6	-1.4	[illegible]
(°F)	(25.0)	(27.0)	(36.1)	(47.5)	(57.9)	(68.0)	(73.8)	(72.3)	(65.7)	(54.3)	(40.3)	(29.5)	(4
PRECIP cm	4.9	4.8	6.8	7.4	9.0	9.3	8.4	8.0	7.6	6.7	5.9	5.0	8
(in.)	(1.9)	(1.9)	(2.7)	(2.9)	(3.5)	(3.7)	(3.3)	(3.1)	(3.0)	(2.6)	(2.3)	(2.0)	(3

Köppen climate classification symbol:________; name:________________________________

____________________; explanation for this determination:________________________________

____

14. Can you see any indication of marine or continental influences based on the two graphs you just completed? Be specific.

____

____

____.

15. Relative to atmospheric temperature, locate a properly installed thermometer either at the college or university you are attending, or perhaps at home. If you do not have access to a thermometer, find another source of information about local temperature: a radio or television station, a local cable channel, the Weather Channel, a local newspaper, the National Weather Service or Environment Canada office.

For at least five days, record the air temperature each day at approximately the same time. See if you can detect a trend or relationship of air temperature to other atmospheric phenomena. Spaces are provided here for you to record your observations.

Air temperature observations (°C and °F):

Day 1:________C°_____F° Place:____________________Time:________

Day 2:________C°_____F° Place:____________________Time:________

Day 3:________C°_____F° Place:____________________Time:________

Day 4:________C°_____F° Place:____________________Time:________

Day 5:________C°_____F° Place:____________________Time:________

16. On a visit to Mount Shasta City (elevation 900 m; 2950 ft) you find the outside air temperature to be 26°C (79 °F), and you also find the daytime weather conditions about average in terms of normal lapse rate (6.4C°/1000m, 3.5F°/1000 ft) in the lower atmosphere. What would the air temperature be (°C and °F) at the summit of Mount Shasta (elevation 4315 m; 14,100 ft) if it is 26°C in the city?

________°C________°F. Show your work:______________________________________

____

17. Using the maps in Chapter 3, analyze the annual temperature pattern experienced in northcentral Siberia. Use specifics from the text and the temperature maps for your response.

____

____

____

____.

18. The wind chill chart in Figure 3-19 (p. 98) presents the apparent temperature that you would experience under different temperature and wind conditions (in °C and °F). Determine the wind-chill temperature for each of the following examples.

a) Wind speed: 24 kmph, air temperature: –34°C = wind chill temp:________________

b) Wind speed: 48 kmph, air temperature: – 7°C = wind chill temp:________________

c) Wind speed: 8 kmph, air temperature: + 4°C = wind chill temp:________________

19. The heat index (HI) is reported for regions that experience high relative humidity and high temperature readings. If the relative humidity is 80% and the air temperature is 32.2°C (90°F) then the National Weather Service heat index (HI) rating is a

Category_____ HI with an apparent temperature of ________________ (in °C and °F).

20. Explain in your own words the six factors that contribute to urban microclimates (pp. 100-02, Figure 3-21, and Table 3-1.

a)____

____

____;

b)____

____

____;

c)____

____

____;

d)____

____

____;

e)____

____

____;

f)____

____

____.

Sample Self-test

(Answers appear at the end of the study guide.)

1. Shortwave and longwave energy passes through the atmosphere or water by
   a) absorption
   b) transmission
   c) refraction
   d) insolation

2. The insolation received at Earth's surface is
   a) usually greatest at the equator
   b) generally greater at high latitudes
   c) is greatest over low-latitude deserts with their cloudless skies
   d) usually lowest along the equator because of cloud cover

3. Infrared energy is mainly absorbed in the "greenhouse effect" by which two gases?
   a) oxygen and hydrogen
   b) ozone and dust
   c) nitrogen and oxygen
   d) water vapor and carbon dioxide

4. The reflective quality of a surface is known as its
   a) absorption
   b) albedo
   c) scattering
   d) wavelength

5. Which of the following has the lowest albedo?
   a) Earth, surface average
   b) fresh snow
   c) grasslands
   d) the Moon's surface in full sunlight

6. The assimilation of radiation by a surface and its conversion from one form to another is termed
   a) reflection
   b) diffuse radiation
   c) absorption
   d) transmission

7. Daily temperatures usually
   a) reach a high at the time of the noon Sun
   b) experience a low in the middle of the night (midnight)
   c) record a high that lags several hours behind the noon Sun
   d) show no relationship to insolation input

8. The quantity of energy mechanically transferred from the air to surface and surface to air is
   a) latent heat energy
   b) sensible heat transfer
   c) ground heating
   d) net radiation

9. The highest albedo value listed in Figure 3-3 is____________ for____________________;

the lowest albedo value listed is ________________for____________________________.

10. List the radiatively active gases that produce Earth's greenhouse.______________________

____.

11. Air temperature is a measure of which of the following in the air?
   a) apparent temperature
   b) relative humidity
   c) sensible heat
   d) heat index

12. Relative to latitude and surface energy receipts, which of the following is true?
   a) insolation intensity decreases with distance from the subsolar point
   b) insolation intensity increases with distance from the subsolar point
   c) daylength is constant across all latitudes
   d) seasonal effects increase toward the equator

13. Relative to temperatures, clouds generally
   a) increase temperature minimums and maximums
   b) are moderating influences in the atmosphere, acting like insulation
   c) cover about 10% of Earth's surface at any one time
   d) decrease nighttime temperatures and increase daytime temperatures

14. In general more moderate temperature patterns
   a) are created by continentality
   b) are exemplified by Wichita, Kansas
   c) indicate maritime influences
   d) occur at altitude

15. The January mean temperature map (Figure 3-15) shows that
   a) isotherms in North America trend poleward
   b) the coldest region on the map is in central Canada
   c) isotherms trend equatorward in the continental interiors of the Northern Hemisphere
   d) the thermal equator is north of the equator

16. Our individual perception of temperature is termed
   a) sensible heat
   b) air temperature
   c) the heat index
   d) apparent temperature

17. The study of climate at or near Earth's surface is called_______________________________.

18. Urban heat islands experience climatic effects related to their
   a) artificial surfaces
   b) lower albedo values
   c) irregular geometric shapes and angles
   d) human occupation and energy conversion systems
   e) all of these are correct

19. Which of the following climatic factors <u>decreases</u> as a result of urbanization
   a) clouds and fog
   b) annual mean temperatures
   c) the presence of condensation nuclei
   d) precipitation
   e) radiation receipts at the surface

20. The blue color of Earth's lower atmosphere is produced by a phenomena known as scattering.
   a) true
   b) false

21. Net radiation at Earth's surface is a product of all incoming shortwave and longwave radiation.
   a) true
   b) false

22. Commercially produced electricity from solar energy conversion has not yet been achieved according to the text.
   a) true
   b) false

23. Sensible heat energy present in the atmosphere is expressed as air temperature.
   a) true
   b) false

24. Air temperature is usually measured with alcohol or mercury thermometers.
   a) true
   b) false

25. Both the Celsius and Fahrenheit temperature scales are in widespread general use worldwide.
   a) true
   b) false

26. Air temperatures, as reported by the National Weather Service or Environment Canada, are always measured in full sunlight.
   a) true
   b) false

27. The greatest annual range of temperatures on Earth occurs in north central Asia.
   a) true
   b) false

28. The temperature at which all motion in a substance stops is called 0° absolute temperature. Its equivalent in different temperature-measuring schemes is __________Celsius (C), ________Fahrenheit (F), and ______________Kelvin. The Fahrenheit scale places the freezing point of water at___________°F(______________°C,____________K) and the boiling point of water at ____________°F (___________°C,____________K). The_____________________is the only major country still using the Fahrenheit scale.

29. The coldest natural temperature ever recorded on Earth was (station, location, date, temperature reading): ____
____.

30. The warmest natural temperature ever recorded on Earth was (station, location, date, temperature reading): ____
____.

Name:______________________ Class Section:______________

Date:______________ Score/Grade:__________

Atmospheric and Oceanic Circulations

Chapter Overview

Earth's atmospheric circulation is an important transfer mechanism for both energy and mass. In the process, the energy imbalance between equatorial surpluses and polar deficits is partly resolved, Earth's weather patterns are generated, and ocean currents are produced. Human-caused pollution also is spread worldwide by this circulation, far from its point of origin. In this chapter we examine the dynamic circulation of Earth's atmosphere that carried Mount Pinatubo's debris worldwide and also carries the everyday ingredients oxygen, carbon dioxide, and water vapor around the globe.

The keys to this chapter are in several integrated figures: the portrayal of winds by the *SEASAT* image (Figure 4-4), the three forces interacting to produce surface pressure wind patterns (Figure 4-10), surface pressure patterns (Figure 4-13), and the synthesis of concepts in the two views of Figure 4-15.

Learning Objectives

The following learning objectives help guide your reading and comprehension efforts. The operative word is in *italics*. Read and work with these carefully and note that exercise #1 asks you about five of these objectives. After reading the chapter and using this workbook, you should be able to:

1. *Discern* the interconnectedness of nations and societies because of Earth's dynamic atmospheric circulation.
2. *Identify* the important global role of the atmospheric circulation in energy and mass balances.
3. *Define* wind. *Describe* how wind is measured, wind direction determined, and how winds are named.
4. *Portray* the nature of air pressure and its decreasing value relative to increasing altitude.
5. *Relate* the story of Evangelista Torricelli and *explain* the instrument that he developed.
6. *Recite* the lowest and highest pressure recorded on Earth, their locations and probable causes.
7. *Differentiate* and *explain* the three driving forces within the atmosphere: pressure gradient, Coriolis, and friction forces.
8. *Define* an isobar and *interpret* an isobaric map.
9. *Explain* the Coriolis force and the apparent deflection created.
10. *Diagram* the vertical and horizontal circulations in a cyclonic and anticyclonic system.

11. *List* the primary high and low pressure areas on Earth and *identify* whether they are produced by dynamic or thermal causes.
12. *Describe* the ITCZ and *relate* it to the equatorial low pressure trough.
13. *Explain* the generation of the trade winds and the westerlies.
14. *Diagram* a simple sketch of a Hadley cell in cross-section.
15. *Portray* the belt of subtropical high pressure and *name* the individual cells.
16. *Relate* the derivation of the terms doldrums and horse latitudes.
17. *Describe* the nature of the polar front and *identify* specific subpolar low-pressure cells.
18. *Describe* upper air circulation and its support role for surface systems.
19. *Define* the jet stream and *differentiate* between polar and subtropical systems.
20. *Explain* several types of local winds: land-sea breezes, mountain-valley breezes, and katabatic winds.
21. *Differentiate* between the winter-dry and summer-wet monsoons of India and Southeast Asia.
22. *Discern* the basic pattern of Earth's major surface and deep ocean currents.

Outline Headings and Glossary Review

These are the first- second-, and third-order headings that divide this chapter. The key terms and concepts that appear **boldface** in the text are listed under their appropriate heading in bold italics; these highlighted terms appear in the text glossary. A check-off box is placed next to each key term so you can mark your progress through the chapter as you define these in your reading notes or prepare note cards.

Wind Essentials

Wind: Description and Measurement
- ❑ ***wind***
- ❑ ***anemometer***
- ❑ ***wind vane***

Global Winds

Air Pressure
- ❑ ***air pressure***
- ❑ ***mercury barometer***
- ❑ ***aneroid barometer***

Driving Forces Within the Atmosphere

- ❑ ***pressure gradient force***
- ❑ ***Coriolis force***
- ❑ ***friction force***

Pressure Gradient Force
- ❑ ***isobars***

Coriolis Force
- ❑ ***geostrophic winds***

Friction Force
- ❑ ***anticyclone***
- ❑ ***cyclone***

Atmospheric Patterns of Motion

Primary High-Pressure and Low-Pressure Areas
- ❑ ***equatorial low-pressure trough***
- ❑ ***polar high-pressure cells***
- ❑ ***subtropical high-pressure cells***
- ❑ ***subpolar low-pressure cells***

Equatorial Low-Pressure Trough
- ❑ ***intertropical convergence zone (ITCZ)***
- ❑ ***Hadley cell***
- ❑ ***trade winds***

Subtropical High-Pressure Cells
- ❑ ***westerlies***

Subpolar Low-Pressure Cells
- ❑ ***polar front***

Polar High-Pressure Cells
- ❑ ***polar easterlies***
- ❑ ***Antarctic high***

Upper Atmospheric Circulation
- ❑ ***Rossby waves***

Jet Streams
- ❑ ***jet stream***

Local Winds
- ❑ ***katabatic winds***

Monsoonal Winds
- ❑ ***monsoon***

Oceanic Currents

Surface Currents
- ❑ ***gyres***
- ❑ ***western intensification***
- ❑ ***equatorial countercurrent***

Deep Currents
- ❑ ***upwelling current***
- ❑ ***downwelling current***

SUMMARY

Learning Activities and Critical Thinking

1. Select any five learning objectives from the list presented at the beginning of this chapter. Place the number selected in the space provided (no need to rewrite each objective). Using the following questions as guidelines only, briefly discuss your treatment of the objective.

- What did you know about the objective before you began?
- What was your plan to complete the objective?
- Which information source did you use in your learning (text, or other)?
- Were you able to complete the action stated in the objective? What did you learn?
- Are there any aspects of the objective about which you want to know more?

a)____:____

____

____

____

____

____.

b)____:____

____

____

____

____

____.

c)____:____

____

____

____

____

____.

d)____:____

____

____

____

____

____.

e)____:____

____

____

____

____

____.

2. Examine Figure 4-1 (a) through (d) (p. 109), read the caption, and the related section of the text (p. 108). Answer the following completion items and questions.

(a) Describe the event depicted in the multiple figure____

____

____.

(b) What remote-sensing mechanisms aboard which satellite made these images?

____

____.

(c) What is "AOT" and how does it appear on the images?____

____

____.

(d) Describe the progression of aerosols between June 15 and August 21, 1991.

____

____

____.

(e) Check the index in ELEMENTAL GEOSYSTEMS and list the other page numbers in the text that present information about the 1991 Mount Pinatubo eruption.________________

____.

3. Relative to wind:

(a) What is wind?____

____.

(b) How are winds named as to their direction?____________________________

____.

(c) What instrument measures wind speed and direction?______________________

____.

4. A descriptive scale useful in visually estimating winds is the traditional *Beaufort wind scale*. Originally established in 1806 by Admiral Beaufort of the British Navy for use at sea, the scale was expanded to include wind speeds by G. C. Simpson in 1926 and standardized by the National Weather Service (Weather Bureau) in 1955. It still is referenced on ocean charts and is presented here with descriptions of visual wind effects on land and sea. This observational scale makes possible the estimation of wind speed without instruments, although most modern ships use sophisticated equipment to perform such measurements.

Even if you do not have access to an anemometer or weather report, you can use the descriptions on the Beaufort scale to estimate winds. You may find this scale useful in the future. Using the Beaufort wind scale presented on the next page (p.64), complete the following table.

Wind Speed			Beaufort Number	Wind Description	Observed Effects at Sea	Observed Effects on Land
kmph	mph	knots				
			1			
			4			
			7			
			9			
			11			

Beaufort Wind Scale (modified and updated)

Wind Speed			Beaufort Number	Wind Description	Observed Effects at Sea	Observed Effects on Land
kmph	mph	knots				
<1	<1	<1	0	Calm	Glassy calm, like a mirror	Calm, no movement of leaves
1–5	1–3	1–3	1	Light air	Small ripples, wavelet scales, no foam on crests	Slight leaf movement, smoke drifts, wind vanes still
6–11	4–7	4–6	2	Light breeze	Small wavelets; glassy look to crests, which do not break	Leaves rustling; wind felt, wind vanes moving
12–19	8–12	7–10	3	Gentle breeze	Large wavelets, dispersed whitecaps as crests break	Leaves and twigs in motion, small flags and banners extended
20–29	13–18	11–16	4	Moderate breeze	Small, longer waves; numerous whitecaps	Small branches moving; raising dust, paper litter,and dry leaves
30–38	19–24	17–21	5	Fresh breeze	Moderate, pronounced waves; many whitecaps; some spray	Small trees and branches swaying, wavelets forming on inland waterways
39–49	25–31	22–27	6	Strong breeze	Large waves, white foam crests everywhere; some spray	Large branches swaying, overhead wires whistling, difficult to control an umbrella
50–61	32–38	28–33	7	Moderate (near) gale	Sea mounding up, foam and sea spray blown in streaks in the direction of the wind	Entire trees moving, difficult to walk into wind
62–74	39–46	34–40	8	Fresh gale (or gale)	Moderately high waves of greater length, breaking crests forming seaspray, well-marked foam streaks	Small branches breaking, difficult to walk, moving automobiles drifting and veering
75–87	47–54	41–47	9	Strong gale	High waves, wave crests tumbling and the sea beginning to roll, visibility reduced by blowing spray	Roof shingles blown away, slight damage to structures, broken branches littering the ground
88–101	55–63	48–55	10	Whole gale (or storm)	Very high waves and heavy, rolling seas; white appearance to foam-covered sea; overhanging waves; visibility reduced	Uprooted and broken trees, structural damage, considerable destruction, seldom occurring
102–116	64–73	56–63	11	Storm (or violent storm)	White foam covering a breaking sea of exceptionally high waves, small and medium-sized ships lost from view in wave troughs, wave crests frothy	Widespread damage to structures and trees, a rare occurrence
>117	>74	>64	12–17	Hurricane	Driving foam and spray filling the air, white sea, visibility poor to nonexistent	Severe to catastrophic damage, devastation to affected society

5. Locate a barometer either at the college or university you are attending, or perhaps at home. Or, find a local weather broadcast on television or radio, a local cable channel, the Weather Channel, a local newspaper, the National Weather Service or Environment Canada that reliably presents barometric pressure information. For at least five days, record the air pressure each day at approximately the same time, if possible. See if you can detect a trend or the relationship between air pressure and other atmospheric phenomena. Spaces are provided here for you to record your observations. Note: you can do approximate conversions using the scale below Figure 4-8, p. 114.

Air pressure observations

Day 1:__________mb__________in. Place:____________________Time:_______

Day 2:__________mb__________in. Place:____________________Time:_______

Day 3:__________mb__________in. Place:____________________Time:_______

Day 4:__________mb__________in. Place:____________________Time:_______

Day 5:__________mb__________in. Place:____________________Time:_______

6. Relative to air pressure:

a) What is normal sea-level pressure expressed in mb and in.?

____.

b) What instruments are used to measure air pressure?____________________

____.

c) Air pressure decreases with altitude (Figure 4-6 and text on p. 112). How much of the atmosphere is compressed below each of the following altitudes?

5500 m (18,000 ft):______________________________

10,700 m (35,100 ft):______________________________

16,000 m (52,500 ft):______________________________

50 km (31 mi):______________________________

d) According to the text, the following records are mentioned for air pressure?

Earth's record low pressure:______________________________

U. S. record low pressure:______________________________

U. S. record high pressure:______________________________

Earth's record high pressure:______________________________

7. Summarize the three driving forces within the atmosphere: briefly define and describe the influence of each:

(a)_____________________:____

____.

(b)_____________________:____

____.

(c)_____________________:____

____.

8. Complete Earth's primary high and low pressure areas from Table 4-1 (p. 121).

Name	**Cause**	**Location**	**Air temperature/ moisture**

9. How many subpolar-low cyclonic systems can you identify in the *Galileo* remote-sensing image in Figure 4-17, p. 126, (surrounding Antarctica)? _____________________.

10. Differentiate between the polar jet stream and the subtropical jet stream using physical descriptions (Figure 4-19 and text pages 126-8). Locate these on the atmospheric cross-section shown in Figure 4-15 (b), p. 123.

(a) Polar jet stream:____

____

____.

(b) Subtropical jet stream:____

____.

11. Describe the effects of the jet stream on weather patterns in the midlatitudes (text pages 126-28). How do jet streams affect flight times (News Report #3, p. 128)?

____

____

____

____.

12. In the space provided, reproduce in a simple diagram the general atmospheric circulation depicted in Figure 4-15(a) and add appropriate labels.

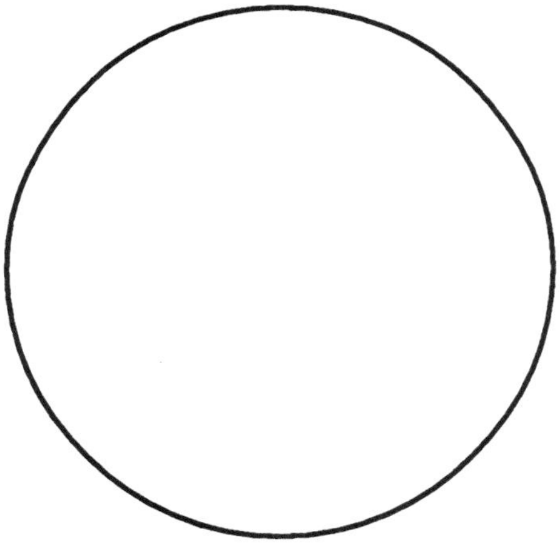

____

13. Now to review, refer back to an image from the *Seasat* satellite that appears in Figure 4-4, p. 111. Answer the following completion items and questions.
(a) Can you identify the northeast and southeast trade winds as they converge along the intertropical convergence zone (ITCZ)? If so, describe the location and pattern on the image.

____

____.

(b) Can you locate a large anticyclonic circulation system in the north Pacific Ocean? If so, describe the location and mention the direction of the wind flow. (See Figure 6-6c.)

____

____

____.

(c) Can you locate a cyclonic (clockwise) circulation in the south Pacific Ocean? If so, describe the location and mention the direction of the wind flow. (Compare this in your answer to the illustration in Figure 4-10(c) and the July pressure map in Figure 4-13(b).

____

____

____

____.

14. Figure 4-22 (a) and (b) (p. 133) presents the Asian monsoonal patterns for January and July. On the following outline version of the illustration, complete the labeling and add directional arrows to indicate wind patterns.

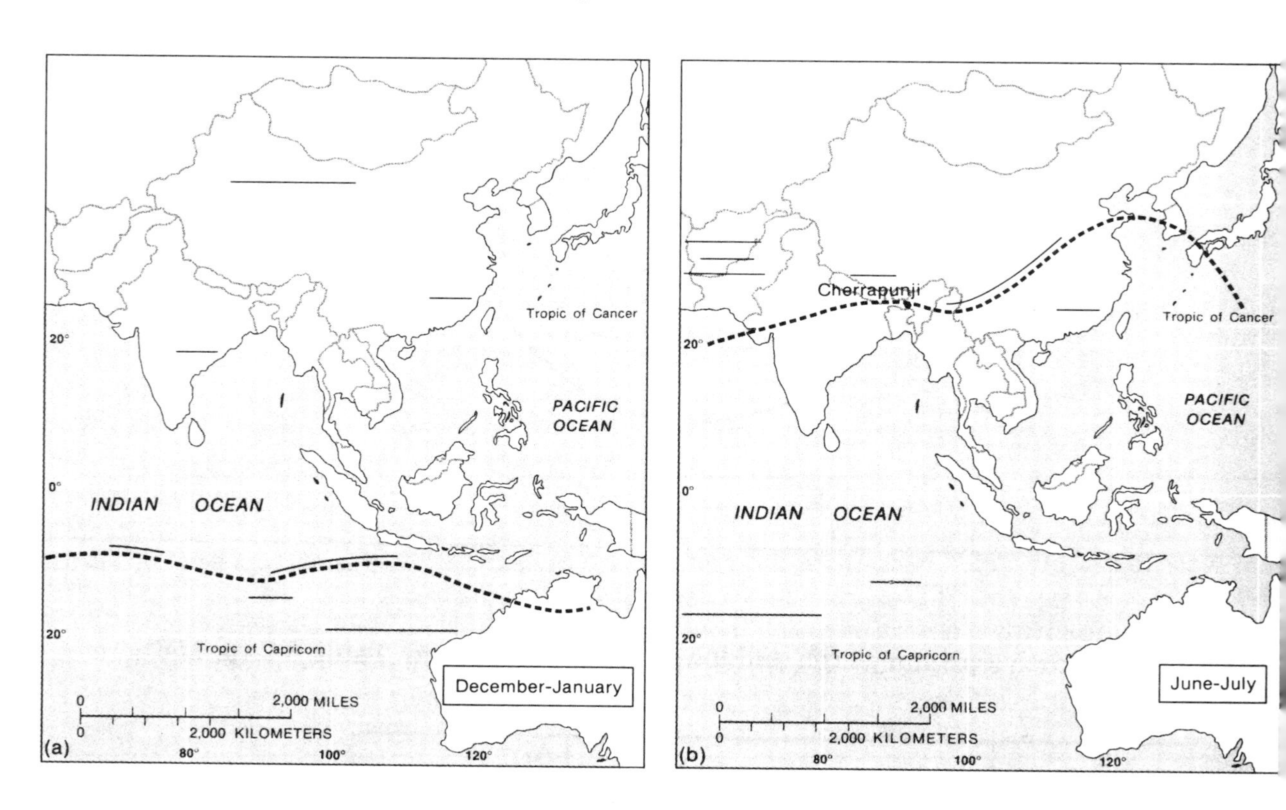

15. Assess the potential for wind-generated electricity (FYI Report 4-1, pages 131-32). What appears to be the major stumbling block to development? Briefly describe present installations (location and amount).

____

____

____

____

____

____

____.

Sample Self-test

(Answers appear at the end of the study guide.)

1. Normal sea level pressure has a value of
   a) 1013.2 millibars or 760 mm of mercury
   b) 28.50 inches of lead or dirt
   c) 32.01 inches of mercury
   d) 506.5 millibars

2. The horizontal motion of air relative to Earth's surface is
   a) barometric pressure
   b) wind
   c) convection flow
   d) an indicator of temperature

3. Which of the following describes the friction force?
   a) drives air from areas of higher to lower barometric pressure
   b) decreases with height above the surface
   c) causes apparent deflection of winds from a straight path
   d) drives air from areas of lower to higher barometric pressure

4. The combined effect of the Coriolis force and the pressure gradient force produces
   a) geostrophic winds
   b) any surface wind
   c) trade winds
   d) air flow moving directly between high and low pressure centers

5. Between 20° to 35° north and 20° to 35° south latitudes, you find
   a) the largest zone of water surpluses in the world
   b) strong westerly winds
   c) the world's arid and semi-arid desert regions and subtropical high pressure
   d) cyclonic systems of low pressure

6. The east side of subtropical high pressure cells (off continental west coasts) tend to be
   a) cool and moist
   b) warm, moist, and unstable
   c) dry, stable, and warm, with cooler ocean currents
   d) generally in the same position all year, not migrating with the high Sun

7. Land-sea breezes are caused by
   a) the fact that water heats and cools faster than land surfaces do
   b) cooler air flowing offshore (toward the ocean) in the afternoon
   c) onshore (toward the land) air flows develop in the afternoon as the land heats faster than the water surfaces
   d) the presence of mountain ranges

8. Monsoonal winds are
   a) regional wind systems that vary seasonally
   b) limited to only the Indian subcontinent
   c) a form of mountain-valley wind
   d) unrelated to the ITCZ position

9. Which of the following is true?
   a) three-fourths of the atmosphere occurs below 10,700 m (35,105 ft)
   b) 90% of the atmosphere is below 5,500 m (18,000 ft)
   c) all but 0.1 % of the atmosphere is accounted for within the troposphere
   d) 90% of the atmosphere remains above the tropopause

10. Air pressure is an expression of the temperature of the atmosphere.
    a) true
    b) false

11. Air pressure is measured with either a mercury or aneroid barometer.
    a) true
    b) false

12. Normal sea level pressure is expressed as 1013.2 mb, or 29.92 in. of Hg, or 760 mm of Hg, or 101.32 kPa.
    a) true
    b) false

13. Atmospheric circulation remains unrelated to the Limited Test Ban Treaty of 1963.
    a) true
    b) false

14. Wind is principally measured with a wind vane and an anemometer.
    a) true
    b) false

15. Winds flow from higher to lower pressure areas as a result of friction force.
   a) true
   b) false

16. Geostrophic winds are <u>surface winds</u> that form in response to the friction force.
   a) true
   b) false

17. A high pressure area is called an anticyclone, a low pressure area a cyclone.
   a) true
   b) false

18. The subtropical belt of high pressure is the place of the intertropical convergence zone.
   a) true
   b) false

19. The principal centers of low pressure in the Northern Hemisphere are the Aleutian and the Icelandic lows associated with the polar front.
   a) true
   b) false

20. The western intensification refers to:____

____

____

____

____.

PART TWO:
Water, Weather, and Climate

Overview–Part Two

Part Two presents spatial aspects of hydrology, meteorology and weather, oceanography, and climate. We begin with water itself—its origin, location, and properties. The dynamics of daily weather phenomena include: the effects of moisture and energy in the atmosphere, the interpretation of cloud forms, conditions of stability or instability, the interaction of airmasses, and the occurrence of violent weather.

The specifics of the hydrologic cycle are explained through the water-balance concept, which is useful in understanding water-resource relationships, whether they are global, regional, or local. Important water resources include rivers, lakes, groundwater, and oceans. The global oceans and seas are identified as the greatest repository of water on Earth. The spatial implications over time of this water-weather system lead to the final topic in Part Two: Earth's climate patterns.

Name:____________________ Class Section:________________

Date:________________ Score/Grade:____________

Atmospheric Water and Weather

Chapter Overview

Water is the essential medium of our daily lives and a principal compound in nature. Water covers 71% of Earth (by area), and within the solar system occurs in such significant quantities only on our planet. Water constitutes nearly 70% of our bodies by weight and is the major ingredient in plants, animals, and our food. A human being can survive 50 to 60 days without food, but only 2 or 3 days without water. The water we use must be adequate in quantity as well as quality for its many tasks. Indeed, water occupies the place between land and sky, mediating energy and shaping both the lithosphere and the atmosphere. We depend on weather to deliver our essential natural water supply.

We begin our study of weather with a discussion of atmospheric stability and instability. We follow huge air masses across North America, observe powerful lifting mechanisms in the atmosphere, examine cyclonic systems, and conclude with a portrait of the violent and dramatic weather that occurs in the atmosphere. Temperature, air pressure, relative humidity, wind speed and direction, daylength, and Sun angle are important measurable elements that contribute to the weather. We tune to a local station for the day's weather report from the National Weather Service in the United States or the Atmospheric Environment Service in Canada to see the current satellite images and to hear tomorrow's forecast.

Learning Objectives

The following learning objectives help guide your reading and comprehension efforts. The operative word is in *italics*. Read and work with these carefully and note that exercise #1 asks you about five of these objectives. After reading the chapter and using this workbook, you should be able to:

1. *Distinguish* Earth from the other planets as the water planet.
2. *Describe* the principal origin of Earth's waters, *relate* the quantity of water that exists today, and *list* the locations of Earth's freshwater supply.
3. *Identify* conditions that can cause worldwide changes in sea level.
4. *Describe* the heat properties of water and *identify* the traits of each phase: solid, liquid, and gas.
5. *Define* latent heat and *explain* the role of this property in nature.
6. *Define* humidity and *explain* the relative humidity concept.
7. *Explain* dew-point temperature and *relate* this to saturated conditions in the atmosphere.
8. *Differentiate* between vapor pressure and specific humidity and *relate* these as expressions of relative humidity.
9. *Explain* the operation of two instruments discussed in the text that are used to measure relative humidity.
10. *Review* the daily pattern of relative humidity and temperature.
11. *List* the measurable elements that contribute to weather.
12. *Define* atmospheric stability and *differentiate* between conditions of stability and instability.
13. *Contrast* the dry adiabatic rate (DAR) and the moist adiabatic rate (MAR) and *relate* them to a parcel of air that is ascending or descending.
14. *Illustrate* unstable and stable atmospheric conditions with a simple graph that relates the environmental lapse rate to the DAR and MAR.
15. *Contrast* and *explain* the two principal processes for raindrop formation and *identify* the necessary requirements for cloud formation.
16. *List* and *recite* the major cloud classes and types.
17. *Contrast* the two basic rain-bearing types of clouds (*nimbo-* and *-nimbus*) and *describe* the nature of associated precipitation.
18. *Identify* the basic types of fog and *explain* the conditions that lead to their formation.
19. *Define* air masses and *relate* them to their source regions, specifically those air masses that affect North America.
20. *Explain* air mass modification and *illustrate* this with an example of a secondary air mass.
21. *Discern* what is an atmospheric lifting mechanism and *list* three principal examples of lifting mechanisms.
22. *Explain* convectional lifting and *portray* it with a specific example. *Relate* this mechanism to the ITCZ.
23. *Explain* orographic lifting and *portray* it with a specific example. *Identify* windward, leeward, and rain shadow concepts, and *review* specific locations where orographic precipitation records were set.

24. *Explain* frontal lifting along the leading edges of contrasting air masses. *Portray* related precipitation, wind, and temperature patterns along a cold front and a warm front.
25. *Describe* the wave-cyclone phenomena and *relate* this to general atmospheric circulation.
26. *Define* cyclogenesis and *outline* the essential stages in the life-cycle of a midlatitude wave cyclone.
27. *Interpret* a daily weather map and *utilize* this tool to determine how basic forecasts are made.
28. *Construct* a diagram of a thunderstorm and *illustrate* the occurrence of thunder, lightning, and hail.
29. *Describe* in simple terms the formation of a mesocyclone and possible tornado development. *Portray* the occurrence of tornadoes across North America.
30. *Outline* tropical cyclone classifications and *describe* hurricanes and typhoons as to their structure, geographical and time-of-year occurrence, and landfall characteristics.
31. *Relate* the tropical events of 1992 and *describe* their impacts on society.

Outline Headings and Glossary Review

These are the first- second-, and third-order headings that divide this chapter. The key terms and concepts that appear **boldface** in the text are listed under their appropriate heading in bold italics; these highlighted terms appear in the text glossary. A check-off box is placed next to each key term so you can mark your progress through the chapter as you define these in your reading notes or prepare note cards.

weather
meteorology

Water on Earth

Quantity Equilibrium

❑ ***outgassing***

Distribution of Earth's Water

Unique Properties of Water

Heat Properties

❑ ***phase change***
❑ ***sublimation***

Ice, the Solid Phase

Water, the Liquid Phase

❑ ***latent heat***

Water Vapor, the Gas Phase

❑ ***latent heat of condensation***

Heat Properties of Water in Nature

❑ ***latent heat of evaporation***

Humidity

❑ ***humidity***

Relative Humidity

❑ ***relative humidity***
❑ ***saturated***
❑ ***dew-point temperature***

Expressions of Relative Humidity

Vapor Pressure

❑ ***vapor pressure***

Specific Humidity

❑ ***specific humidity***

Instruments for Measurement

Atmospheric Stability

Adiabatic Processes

❑ ***normal lapse rate***
❑ ***environmental lapse rate***
❑ ***adiabatic***

Dry Adiabatic Rate (DAR)

❑ ***dry adiabatic rate***

Moist Adiabatic Rate (MAR)

❑ ***moist adiabatic rate***

Stable and Unstable Atmospheric Conditions

Clouds and Fog

❑ ***cloud***
❑ ***fog***
❑ ***condensation nuclei***

Cloud Types and Identification

❑ ***stratus***
❑ ***cumulus***
❑ ***nimbostratus***
❑ ***stratocumulus***
❑ ***cirrus***
❑ ***cumulonimbus***

Fog

❑ ***advection fog***

❑ ***upslope fog***
❑ ***valley fog***
❑ ***evaporation fog***
❑ ***radiation fog***

Air Masses

❑ ***air mass***

Atmospheric Lifting Mechanisms

Convectional Lifting
Orographic Lifting
❑ ***orographic lifting***
❑ ***rain shadow***
Frontal Lifting
❑ ***front***
Cold Front
Warm Front

Midlatitude Cyclonic Systems

❑ ***wave cyclone***
Life Cycle of a Midlatitude Cyclone
❑ ***occluded front***
Daily Weather Map and the Midlatitude Cyclone

Violent Weather

Thunderstorms
Lightning and Thunder
❑ ***lightning***
Hail
❑ ***hail***
Tornadoes
❑ ***mesocyclone***
❑ ***funnel clouds***
❑ ***tornado***
❑ ***waterspout***
Tropical Cyclones
❑ ***tropical cyclone***
❑ ***hurricane***
❑ ***typhoon***
Hurricanes and Typhoons

SUMMARY

Learning Activities and Critical Thinking

1. Select any five learning objectives from the list presented at the beginning of this chapter. Place the number selected in the space provided (no need to rewrite each objective). Using the following questions as guidelines only, briefly discuss your treatment of the objective.

- What did you know about the objective before you began?
- What was your plan to complete the objective?
- Which information source did you use in your learning (text, or other)?
- Were you able to complete the action stated in the objective? What did you learn?
- Are there any aspects of the objective about which you want to know more?

a)____:____

____

____

____

____

____.

b)____ :____

____

____

____

____

____.

c)____ :____

____

____

____

____

____.

d)____ :____

____

____

____

____

____.

e)____ :____

____

____

____

____

____.

2. If deprived of food and water, describe how long a human can survive.

____

____

____.

3. How much water constitutes Earth's hydrosphere (km^3 and mi^3)?

____.

4. The seven largest lakes in the world in terms of volume represent ______% of Earth's freshwater. In one lake, ____________________, some ______% of Earth's freshwater is found.

5. What percentage of freshwater is represented by

Ice sheets and glaciers? ____________________

Freshwater lakes?____________________

Saline lakes and inland seas?____________________

Atmosphere?____________________

Rivers and streams?____________________

All groundwater?____________________

Soil moisture storage?____________________

6. *Name* the term that describes the following phase changes and *list* the latent heat energy (in calories) that is either absorbed or released for one gram making the change in state. (See Figures 5-3 and 5-5, p. 145 and 147.)

(a) solid to liquid (at 0°C):____________________

(b) liquid to vapor (at 20°C):____________________

(c) liquid to solid (at 0°C):____________________

(d) vapor to liquid (at 100°C):____________________

7. So what's with roads, pipes, and sinking ships (News Report #1, p. 148)?____________

____

____

____.

8. Describe two expressions of relative humidity.

(a)____________________

(b)____________________

9. What happens when the *dew-point temperature* and the *air temperature* reach the same temperature? Begin by defining dew-point temperature.

____

____

____

____.

10. Describe the operation of two instruments used to measure relative humidity.

(a)____

____;

(b)____

____.

11. A parcel of air is at 20°C (68°F) and has a saturation vapor pressure of 24 mb. If the water vapor content actually present is exerting a vapor pressure of only 12 mb in 20°C air, the relative humidity is ______________________________. What would the approximate relative humidity be if the parcel of air increased in temperature to 30°C (86°F) (12mb ÷ 40 mb)? ______________________________. What would the approximate relative humidity be if the parcel of air decreased in temperature to 10°C (50°F)?______________________________. Use Figure 5-7, p. 150, in preparing your answers.

12. Using the graph in Figure 5-6, p. 149, *contrast* and *compare* the mean relative humidity and air temperature for a series of typical days in Sacramento, CA..

(a) Relative humidity:____

____.

(b) Air temperature:____

____.

(c) What is the relationship between these two variables shown on the graph?

____

____

____.

13. What are the three principal lifting mechanisms that affect air masses?

(a)____

(b)____

(c)____

14. Plot data lines on this graph that denote the *dry adiabatic rate* (*DAR*) and *moist adiabatic* (*MAR*) rates (see Figure 5-10, p. 153). Label each line. Label the portions of the graph that would represent atmospheric conditions of: **a)** unstable, **b)** conditionally unstable, and **c)** stable—given different values for the *environmental lapse rate*.

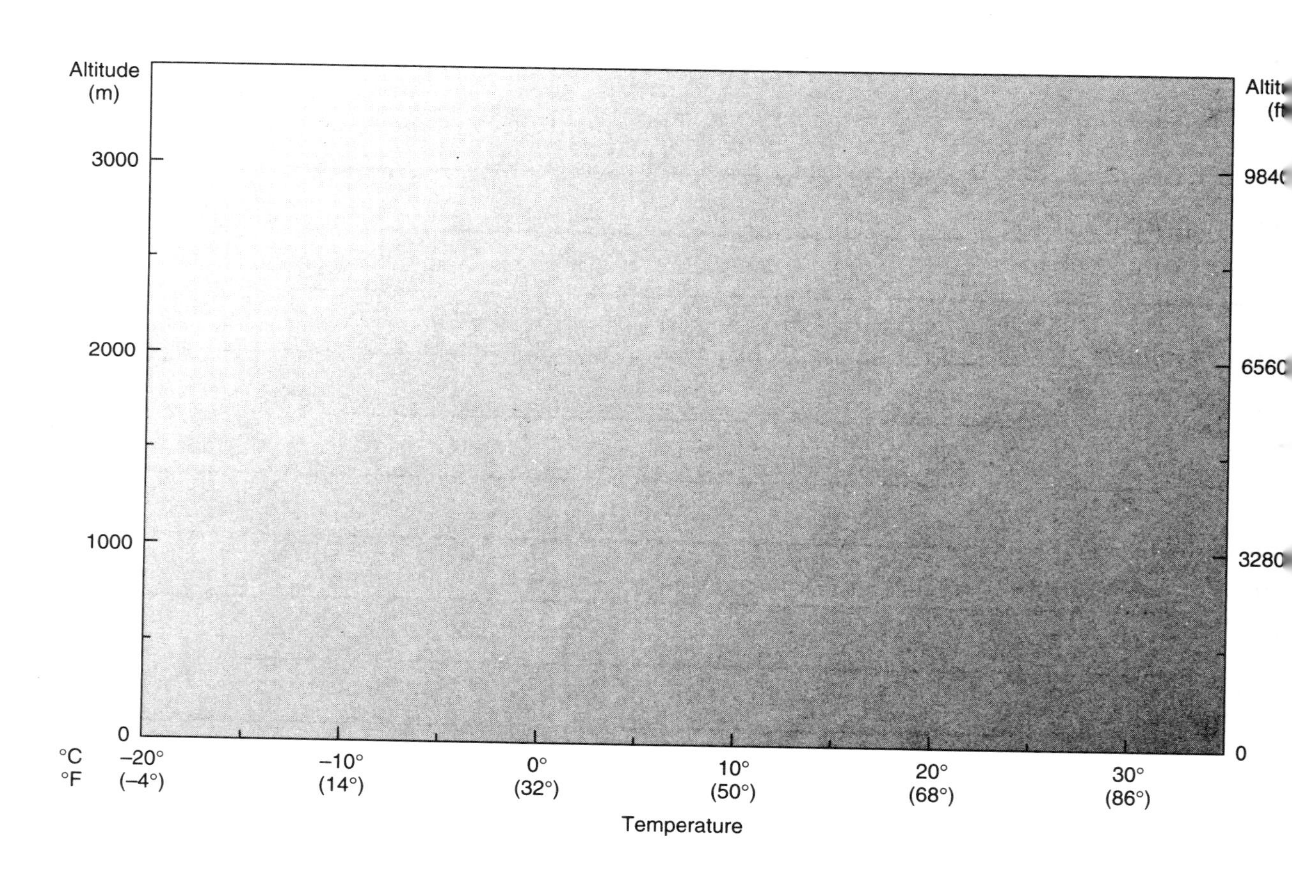

15. Describe atmospheric conditions of stability (unstable and stable) portrayed in the following two graphs.

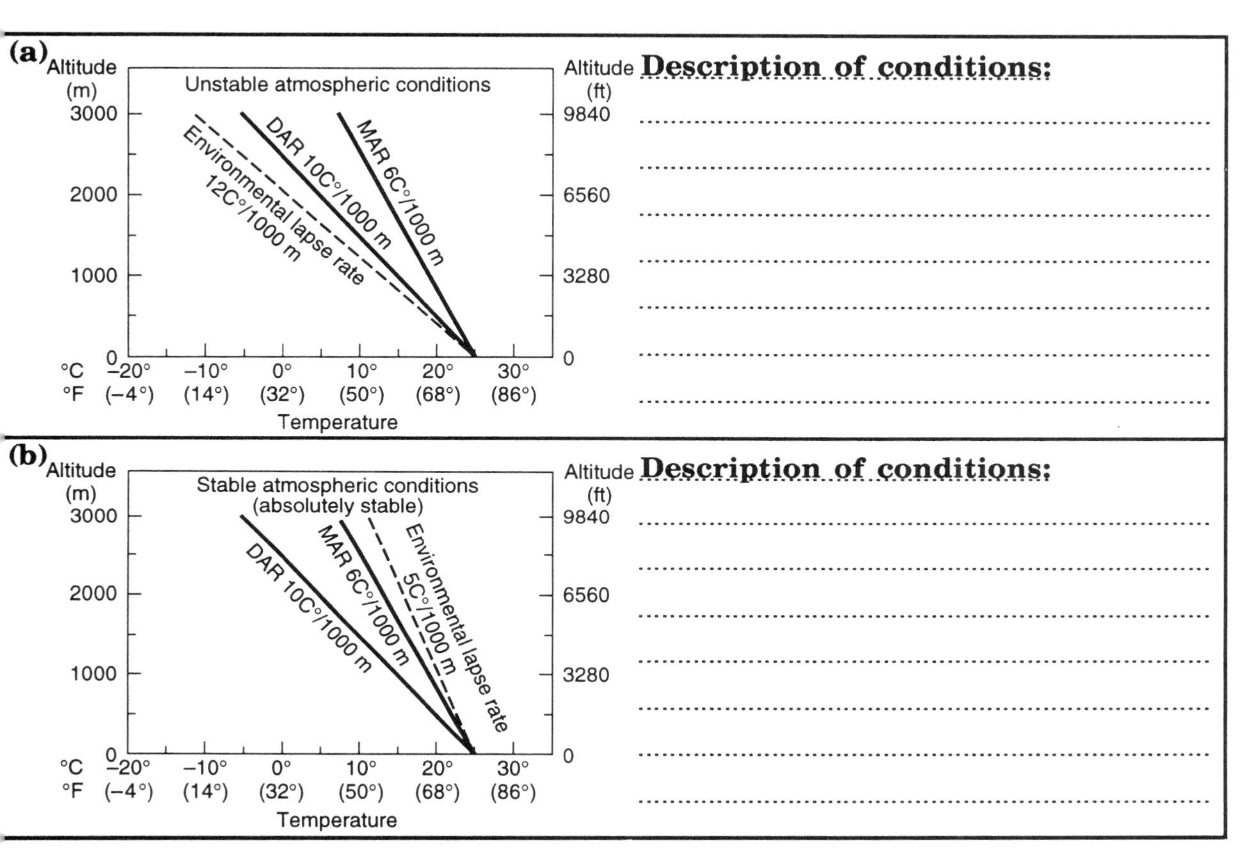

16. Using Figures 5-13, 5-14, and text pp. 154-60, describe the physical characteristics of the cloud types listed below:

Cloud type **Description**

a) stratus: ______________________________

____

b) cumulus: ______________________________

____

c) nimbostratus: ______________________________

____

d) cumulonimbus: ______________________________

____

e) altostratus: ______________________________

____

f) cirrus: ____
____

g) fog: ____
____

17. Name several areas in the United States and Canada that experience the highest annual number of days with heavy fog.

____
____.

18. Using Table 5-1 (p. 162), identify and describe the principal air masses that influence North America on these two maps from Figure 5-18 (p. 162).

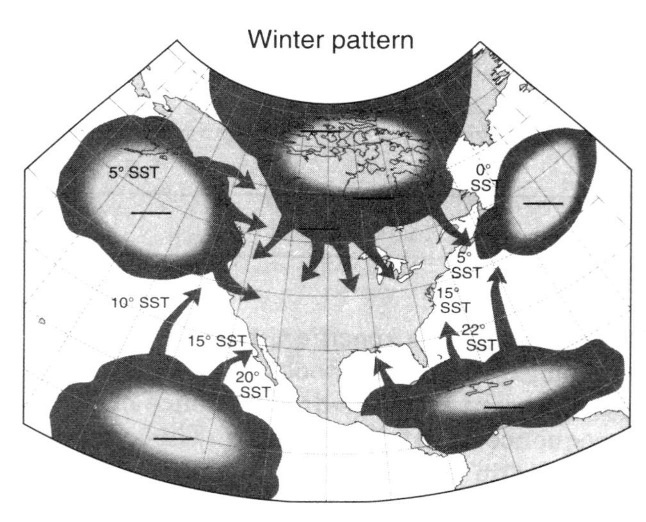

SST = Sea-surface temperature (°C)

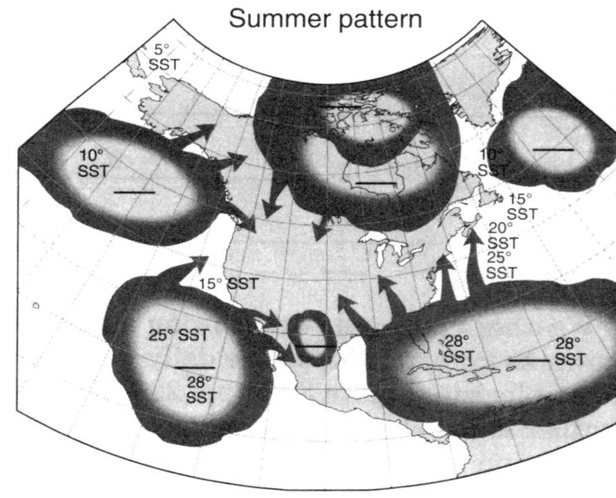

SST = Sea-surface temperature (°C)

19. a) Describe the place that is the wettest *average annual* precipitation on Earth and he location and physical characteristics that produce this total.____________________.

b) Describe the wettest place on Earth that holds the record for a *single year* and the ocation and physical characteristics that produce this total.____________________.

20. Compare and contrast the average characteristics you would experience with the passage of a *cold front* and a *warm front* in North America:

a) Cold front passage—before:____

____;

during:____

____;

after:____

____.

b) Warm front–during:____

____.

21. Weather maps used for showing atmospheric conditions at a specific time and place re known as *synoptic maps*. The daily weather map is a key analytical tool for neteorologists. Figure 5-28 (p. 171) in the text presents a weather map along with a natching satellite image. The symbols and notations used on these maps are noted in he margin of page 171.

Using an adapted version of a synoptic map for April 1, 1971, 7:00 A.M. E.S.T., this ctivity involves adding appropriate isobars, fronts, and air mass designations to the nap. The following discussion takes you step-by-step through the completion of this nap on page 85. A completed version of this weather map is in the answer key (p. 247).

(a) Begin by examining weather conditions on the map. Each circle represents a veather station. The state of the sky is assessed by the degree to which the circle is haded. The two sets of numbers to the upper left of the circle are air temperature upper) and dew-point temperature (lower).

Determine the pattern of low-pressure and high-pressure systems on the map and elated air masses to determine the position of the fronts. The southeastern portion of he country is influenced by a mild maritime tropical (mT) air mass. The high-pressure rea is under the influence of a continental polar (cP) air mass. Note the southerly vinds moving northward toward the warm front that stretches from the center of low pressure in Wisconsin, across Michigan, and into Pennsylvania.

The number to the upper right of the station is the barometric pressure presented in an abbreviated form used by the National Weather Service. As an example: if a *145* appears, it is short for 1014.5 mb; a *980* is 998.0 mb. On April 1, 1971, the center of low pressure was near Wausau, Wisconsin (see the **L** on the map), with a pressure of 994.7 mb (947 on the map). Around this center of low pressure the wind flags at the various stations show counterclockwise winds as you expect in a midlatitude cyclonic circulation system. The high-pressure center is located near Salmon, Idaho (see the **H** on the map), with a pressure of 1033.6 mb (336 on the map). Note the pattern of temperatures associated with a cold-air mass centered in the region of high pressure.

(b) To draw the cold front and leading edge of the cold air mass, locate stations with northwesterly winds and lower temperatures. Now locate cities with southerly and southwesterly winds and higher temperatures (for instance, compare northeastern Missouri with southwestern Indiana, or Arkansas with Oklahoma). Draw the cold front between these contrasting pairs of stations southwestward from Wisconsin to Texas. Use the proper cold-front symbol. Sketch in the warm front stretching to the east from the low-pressure center. Use the proper warm-front symbol.

(c) On the weather map, using a pencil, draw in the isobars connecting points of equal barometric pressure. Begin with **996 mb** around the 947 low-pressure center (pressures lower than 996 go inside the isobar, higher than 996 go outside this closed isobar). Now draw in order the rest of the isobars at 4 mb intervals: **1000**, **1004**, **1008**, **1012**, **1016**, **1020**, **1024**, **1028**, and **1032 millibars**. Note the 1032 mb isobar is a closed isobar that surrounds the Salmon, Idaho, region. A completed version of this map is presented for your reference in the "Answer Key to Self-tests" section in the back of this study guide. Please complete the weather map analysis before consulting this key.

(d) Recall that the relationship between air temperature and dew-point temperature gives you an idea of the moisture content of the cP (dry) and mT (moist) air masses, respectively. List a sampling of air temperatures and dew-point temperatures for Idaho, Wyoming, and Utah:

Idaho:____

Wyoming:______________________________________

Utah:____

(e) Now compare these with air temperatures and dew-point temperatures in Louisiana and Mississippi:

Louisiana:_____________________________________

Mississippi:___________________________________

(f) Describe the pattern of cloudiness across the map. The pattern of cloudiness is indicated by the "state-of-the-sky" status recorded within each station symbol. The areas of frontal lifting are clearly identified by these patterns of clouds.

____

____

____.

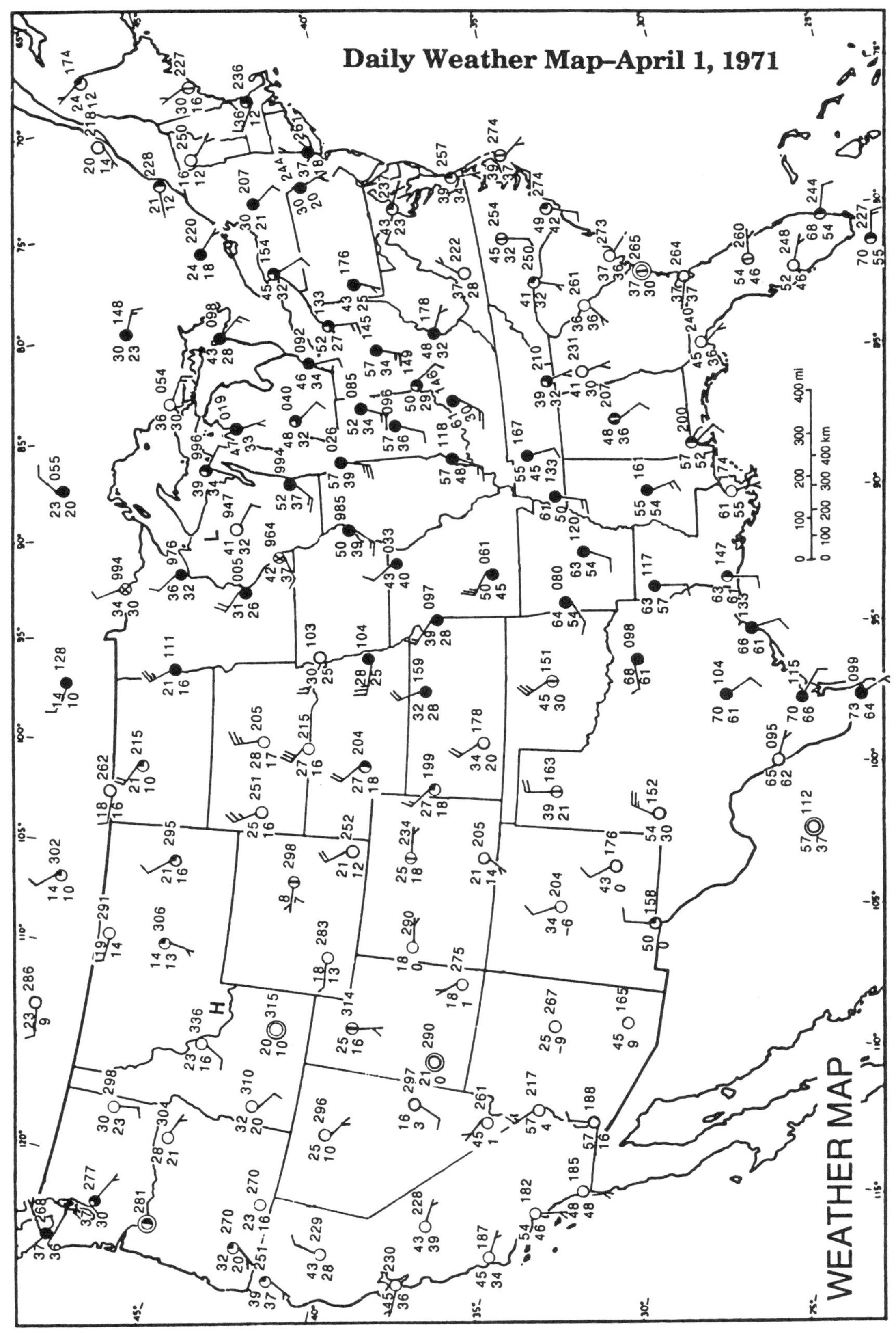
Daily Weather Map–April 1, 1971
WEATHER MAP
0 100 200 300 400 mi
0 100 200 300 400 km

22. Describe a thunderstorm (cloud type and precipitation, thunder, lightning, and hail potential).____

____

____

____

____.

23. For the period 1959-1993, list the three top months for tornadoes:

(a)____________________

(b)____________________

(c)____________________

24. In terms of hazard perception and the lessons learned from Hurricane Camille and Hurricane Andrew, please assess the events of 1992 and the impact of this event on society (pp. 177-81).

____

____

____

____.

Sample Self-test

(Answers appear at the end of the study guide.)

1. Water covers approximately what percentage of Earth's surface?
   - a) 50%
   - b) 71%
   - c) 83%
   - d) less than 50%

2. The present quantity of water on Earth, functioning in the hydrologic cycle, was achieved approximately____________________ years ago, according to the text.
   - a) one million years ago
   - b) one billion years ago
   - c) two billion years ago
   - d) since the last ice age

3. Water has unusual heat properties, partially related to hydrogen bonding, when compared with other compounds. If you were to take one gram of ice at 0°C and raise it to one gram of water vapor at 100°C, how many total calories would you need to add in terms of latent and sensible heat?

a) 540 calories
b) 80 calories
c) 320,000 calories per square centimeter
d) 720 calories

4. The major portion of freshwater today is located in

a) all sub-surface water
b) groundwater resources
c) ice sheets and glaciers
d) the major rivers and lakes and atmospheric moisture

5. Relative humidity refers to

a) the amount of water vapor in the air compared with normal levels
b) the amount of moisture in the air relative to your own sensible feelings
c) the actual humidity in the air, or the absolute humidity
d) the amount of water vapor in the air at a given temperature and pressure, expressed as a percentage of the moisture capacity of the air

6. The principal process of raindrop formation in tropical warm clouds is

a) a collision-coalescence process
b) condensation nuclei
c) an ice-crystal process
d) supercooled droplets

7. Clouds that have strong vertical development and produce precipitation are called

a) stratocumulus
b) cumulonimbus
c) nimbostratus
d) cumulus

8. Assume that a warm air bubble, or parcel, near Earth's surface, that has a temperature of 25°C, begins to rise. Assume that the parcel of air contains 10 mb vapor pressure. At what altitude will the lifting mass of air become saturated? (Use the saturation vapor pressure graph in Figure 5-7, p. 150, and a DAR of 1C° per 100 m).

a) 400 m
b) 1000 m
c) 1500 m (15C° of cooling to 10°C)
d) it does not reach the dew-point temperature in this example

9. Which of the following weather conditions best describes your area today?

a) dominance over the area by a high pressure cell
b) a modified cP air mass
c) the passage of a midlatitude wave cyclone
d) local heating, creating convection and thundershowers

10. The dry adiabatic rate (DAR) is
   a) 6C° per 1000 m (3.3F° per 1000 ft)
   b) the rate used in a lifting saturated parcel of air
   c) the environmental lapse rate
   d) 10C° per 1000 m (5.5F° per 1000 ft)

11. Maritime tropical Pacific (mT) air and maritime tropical Gulf and Atlantic (mT) air
   a) are very different from one another, since they occur over cool and warm ocean surfaces, respectively
   b) are identical in strength, since both carry the same mT destination
   c) are both usually present during the coldest weather conditions in the east and midwest
   d) are not related to weather

12. Which cloud type is specifically a good indicator of an arriving storm, say within the next 24 hours?
   a) fog
   b) cumulus
   c) stratocumulus
   d) cirrus

13. The wettest average annual place on Earth is located
   a) in the United States
   b) in the Amazon in Brazil
   c) on the slopes of the Himalayas in India
   d) the Monsoons in Southeast Asia

14. With respect to the three main lifting (cooling) mechanisms, which are local heating, orographic, and frontal, which of the following is correct?
   a) the place with the wettest average annual rainfall on Earth is most closely related to local heating and frontal activity
   b) a single convectional storm triggered by local heating affects large geographical regions
   c) we do not get all three mechanisms within the United States
   d) given the necessary physical requirements, orographic precipitation is usually the most consistent type of the three

15. Tornado development is associated with
   a) warm fronts
   b) mesocyclone circulation and cold fronts
   c) continental tropical air masses
   d) stable air masses

16. Afternoon thundershowers in the southeastern United States are more than likely a result of
   a) convectional lifting
   b) orographic lifting
   c) frontal lifting
   d) subtropical high pressure disturbance

17. Earth, like the other planets in the solar system, possesses large quantities of water.
a) true
b) false

18. Glacio-eustatic factors specifically relate to changes in sea level caused by actual physical changes in the elevation of landmasses.
a) true
b) false

19. The largest lake in the world in terms of volume is Lake Superior.
a) true
b) false

20. The phase change of water to ice is called freezing.
a) true
b) false

21. The energy involved in the phase changes of water is called latent heat.
a) true
b) false

22. Relative humidity is the amount of water vapor in the air at a given temperature and pressure, expressed as a percentage of the moisture capacity of the air.
a) true
b) false

23. Clouds are initially composed of raindrops.
a) true
b) false

24. The ______________________________ uses the principle that human hair changes as much as 4% n length between 0 and 100% relative humidity. The ______________________________ has two thermometers mounted side-by-side on a metal holder. That portion of total air pressure that s made up of water vapor molecules is termed ______________________________ and is expressed in millibars (mb).

25. A cloud in contact with the ground is commonly referred to as ____________________. By nternational agreement, this is officially described as a cloud layer on the ground, with visibility restricted to less than ______________________________.
An ______________________________ fog forms when air in one place migrates to another place where saturated conditions exist. An ______________________________ fog, or steam fog, may form as the water molecules evaporate from the water surface into the cold overlying air,

effectively humidifying the air. Because cool air is denser, it settles in low-lying areas, producing an ________________________________ fog in the chilled, saturated layer near the ground. A ____ fog forms when radiative cooling of a surface chills the air layer directly above that surface to the dew-point temperature, creating saturated conditions and fog.

26. Stability refers to the tendency of a parcel of air to either remain as it is or change its initial position.
    a) true
    b) false

27. The moist adiabatic rate (MAR) is used in describing a moving air parcel that is less than saturated.
    a) true
    b) false

28. The world's rainfall records are associated with orographic precipitation.
    a) true
    b) false

29. A cold front is characterized by drizzly showers of long duration.
    a) true
    b) false

30. Cyclogenesis refers to the birth or strengthening of a wave cyclone.
    a) true
    b) false

31. Atmospheric <u>pressure</u> is portrayed on the daily weather map with a pattern of <u>isotherms</u>.
    a) true
    b) false

32. Thunder is produced by the sound of rapidly expanding air heated by lightning.
    a) true
    b) false

33. Typhoons and hurricanes are significantly different types of storms in terms of physical structure.
    a) true
    b) false

34. Relative to orographic precipitation, the wetter intercepting slope is termed the ____________________, as opposed to the drier far-side slope, known as the ____________________.

Name:______________________ Class Section:______________

Date:______________ Score/Grade:____________

Water Resources

Chapter Overview

Water is not always naturally available when and where it is needed. From the maintenance of a house plant to the distribution of local water supplies, from an irrigation program on a farm to the rearrangement of river flows—all involve aspects of water balance and water-resource management. The availability of water is controlled by atmospheric processes and climate patterns.

This chapter begins by examining the water balance, which is an accounting of the hydrologic cycle for a specific area, with emphasis on plants and soil moisture. The nature of groundwater is discussed and several examples are given of this abused resource. Groundwater resources are closely tied to surface-water budgets. Of course, the ultimate repository of water on Earth is the ocean. We also consider the water we withdraw and consume from available resources, in terms of both quantity and quality. Many aspects of this chapter may prove useful to you since we all interact with the hydrologic cycle on a daily basis.

Learning Objectives

The following learning objectives help guide your reading and comprehension efforts. The operative word is in *italics*. Read and work with these carefully and note that exercise #1 asks you about five of these objectives. After reading the chapter and using this workbook, you should be able to:

1. *Relate* the importance of the water-balance concept to your understanding of the hydrologic cycle, water resources, and soil moisture considerations at a site.
2. *Illustrate* the hydrologic cycle with a simple sketch, and *label* each of the components.
3. *Identify* the pathways for precipitation to Earth's surface.
4. *Construct* the water balance equation for the accounting of water supply and *identify* each component (terms only).
5. *Describe* precipitation and a method for its measurement.
6. *Define* potential evapotranspiration.
7. *Explain* empirical measures and estimating procedures for determining potential evapotranspiration.
8. *Define* deficit, *relate* the concept to potential evapotranspiration and *derive* actual evapotranspiration.
9. *Define* surplus and *describe* the flow paths of excess water.
10. *Review* the types of soil moisture and *explain* the role of each in the water balance.
11. *Differentiate* between soil moisture utilization and soil moisture recharge.
12. *Relate* the level of soil moisture to plant efficiency in obtaining needed water.
13. *Relate* the Snowy Mountain Scheme to water balance considerations.

14. *Portray* groundwater resources for Earth as a whole and the United States and Canada in particular.
15. *Describe* the nature of groundwater and *define* the underground structure of the groundwater environment.
16. *Explain* groundwater utilization and *discern* the impact of groundwater mining.
17. *Analyze* the Ogallala Aquifer of the High Plains and *outline* the emerging problem related to that resource.
18. *Contrast* groundwater pollution with surface contamination as to the permanence of pollution.
19. *Define* stream discharge and *express* how it is stated in different types of measurement units.
20. *Explain* the concept of an exotic stream using the Nile and Colorado rivers.
21. *List* the five largest rivers on Earth in terms of discharge at their mouths.
22. *Discuss* the daily water budget for the 48 contiguous states.
23. *Contrast* consumptive uses of water with water withdrawal.
24. *Discern* critical aspects of freshwater supplies for society in the future and *cite* specific related issues.
25. *Analyze* the salinity of the ocean and *differentiate* between brine and brackish regions of the sea.
26. *Identify* the physical structure of the ocean, including temperature, salinity, and dissolved carbon dioxide and oxygen levels.

Outline Headings and Glossary Review

These are the first- second-, and third-order headings that divide this chapter. The key terms and concepts that appear **boldface** in the text are listed under their appropriate heading in bold italics; these highlighted terms appear in the text glossary. A check-off box is placed next to each key term so you can mark your progress through the chapter as you define these in your reading notes or prepare note cards.

The Hydrologic Cycle

- ❑ ***hydrologic cycle***

A Hydrologic Cycle Model

- ❑ ***infiltration***
- ❑ ***percolation***

The Water-Balance Concept

The Water-Balance Equation

Precipitation

- ❑ ***precipitation***
- ❑ ***rain gauge***

Potential Evapotranspiration

- ❑ ***evaporation***
- ❑ ***transpiration***
- ❑ ***evapotranspiration***
- ❑ ***potential evapotranspiration***

Determining POTET

- ❑ ***evaporation pan***

Deficit

- ❑ ***deficit***
- ❑ ***actual evapotranspiration***

Surplus

- ❑ ***surplus***
- ❑ ***total runoff***

Soil Moisture Storage

- ❑ ***soil moisture storage***
- ❑ ***wilting point***
- ❑ ***capillary water***
- ❑ ***available water***
- ❑ ***field capacity***
- ❑ ***gravitational water***
- ❑ ***soil moisture utilization***
- ❑ ***soil moisture recharge***

Three Examples of Water Balances

Water Balance and Water Resources

Groundwater Resources

Groundwater Description

- ❑ ***zone of aeration***
- ❑ ***zone of saturation***
- ❑ ***porosity***
- ❑ ***permeable***
- ❑ ***impermeable***
- ❑ ***aquifer***
- ❑ ***aquiclude***
- ❑ ***unconfined aquifer***
- ❑ ***water table***
- ❑ ***confined aquifer***
- ❑ ***aquifer recharge area***
- ❑ ***artesian water***

Groundwater Utilization

- ❑ ***drawdown***

❑ ***cone of depression***
Pollution of the Groundwater Resource

Distribution of Streams

❑ ***discharge***
❑ ***exotic stream***
❑ ***internal discharge***

Our Water Supply

Daily Water Budget
World Water Economy

Global Oceans and Seas

Salinity and Composition
❑ ***salinity***

SUMMARY

Learning Activities and Critical Thinking

1. Select any five learning objectives from the list presented at the beginning of this chapter. Place the number selected in the space provided (no need to rewrite each objective). Using the following questions as guidelines only, briefly discuss your treatment of the objective.

- What did you know about the objective before you began?
- What was your plan to complete the objective?
- Which information source did you use in your learning (text, or other)?
- Were you able to complete the action stated in the objective? What did you learn?
- Are there any aspects of the objective about which you want to know more?

a)_____:____

____

____

____

____

____.

b)_____:____

____

____

____

____

____.

c)_____:____

____

____

____

____

____.

d)____ :____

____

____

____

____

____.

e)____ :____

____

____

____

____

____.

2. Using Figure 6-1 (p. 187) and the section in the text describing the hydrologic cycle, complete the information and labels on the following outline drawing.

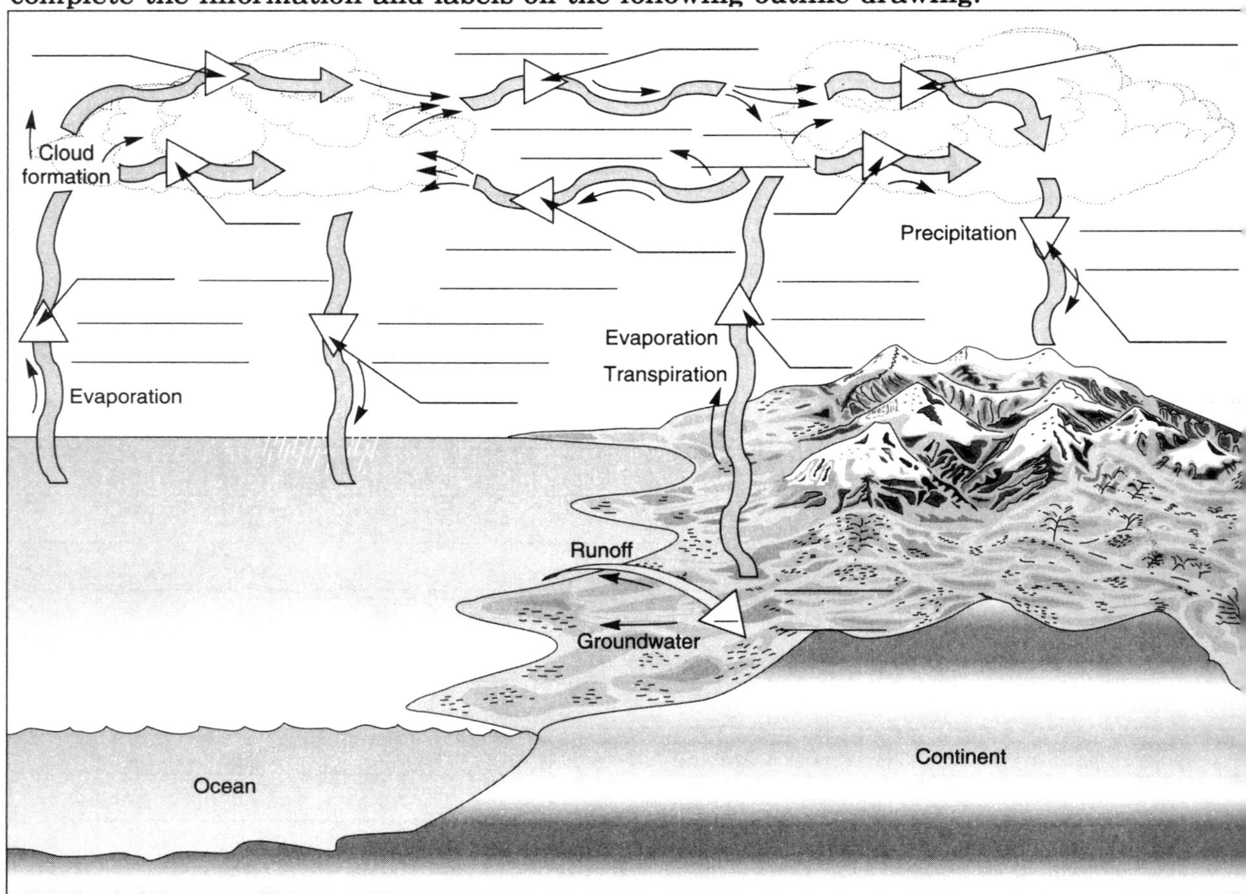

3. What is the purpose of budgeting water and, specifically, the water balance? Is there a usefulness to this concept in your opinion?

____

____

____

____.

4. Annual precipitation (PRECIP, water supply) for the United States and Canada is shown in Figure 6-4 (p. 190), annual potential evapotranspiration (POTET, water demand) is in Figure 6-5 (p. 191). Briefly compare these two maps and answer the following items.

(a) Can you identify from the two maps regions where PRECIP is *higher* than POTET? Describe.

____

____

____.

(b) Can you identify from the two maps regions where POTET is *higher* than PRECIP? Describe.

____

____

____.

(c) Why do you think 95% of irrigated agriculture in the United States occurs west of the 95th meridian (central Kansas)?

____

____.

(d) Where you live, is the natural water demand usually met by the natural precipitation supply? Or, does your region experience a natural shortage? Are there some months of surplus and some months of deficit in the annual pattern of water balance components where you live?.

____

____

____.

5. (a) Record the components of the water balance equation on the following line using the acronyms given in Figure 6-2, p. 188..

______________ = (______________ – ______________) + ______________ ± Δ ______________

(b) Define each of the following components of the water balance equation:

PRECIP:____

____

POTET:____

____

DEFIC:____

____

SURPL:____

____

± Δ STRGE:____

____

ACTET:____

____

6. Briefly relate what happens to plant moisture needs and soil moisture as available water in the soil is reduced by soil moisture utilization (p. 192-93):______________________

____

____

Water balance data for Kingsport, TN, to use in item #7.

Kingsport, Tennessee (Cfa): pop. 32,000, lat. 36° 30′ N, long. 82° 30′ W, elev. 391 m (1284 ft)

	Jan	Feb	Mar	Apr	May	Jun	Jul	Aug	Sep	Oct	Nov	Dec	Annual
Temperature°C	4.3	4.8	8.2	14.1	18.6	23.0	24.6	23.9	21.3	15.0	8.5	4.5	14.2
(°F)	(39.7)	(40.6)	(46.8)	(57.4)	(65.5)	(73.4)	(76.3)	(75.0)	(70.3)	(59.0)	(47.3)	(40.1)	(57.6)
PRECIP cm	9.7	9.9	9.7	8.4	10.4	9.7	13.2	11.2	6.6	6.6	6.6	9.9	111.9
(in.)	(3.8)	(3.9)	(3.8)	(3.3)	(4.1)	(3.8)	(5.2)	(4.4)	(2.6)	(2.6)	(2.6)	(3.9)	(44.1)
POTET cm	0.7	0.8	2.4	5.7	9.7	13.2	15.0	13.3	9.9	5.5	1.2	0.7	78.1
(in.)	(0.3)	(0.3)	(0.9)	(2.2)	(3.8)	(5.2)	(5.9)	(5.2)	(3.9)	(2.2)	(0.5)	(0.3)	(30.7)

7. Using the water-balance graph below, plot the Kingsport, Tennessee, data given on page 96. Use a <u>line graph</u> for precipitation and a <u>dashed line</u> for potential evapotranspiration (you may use different color pencils for these lines). Next, interpret the relationship between these demand and supply concepts: surplus, soil moisture utilization, deficit, and soil moisture recharge. Use the same color designations as in Figure 6-8 (p. 193).

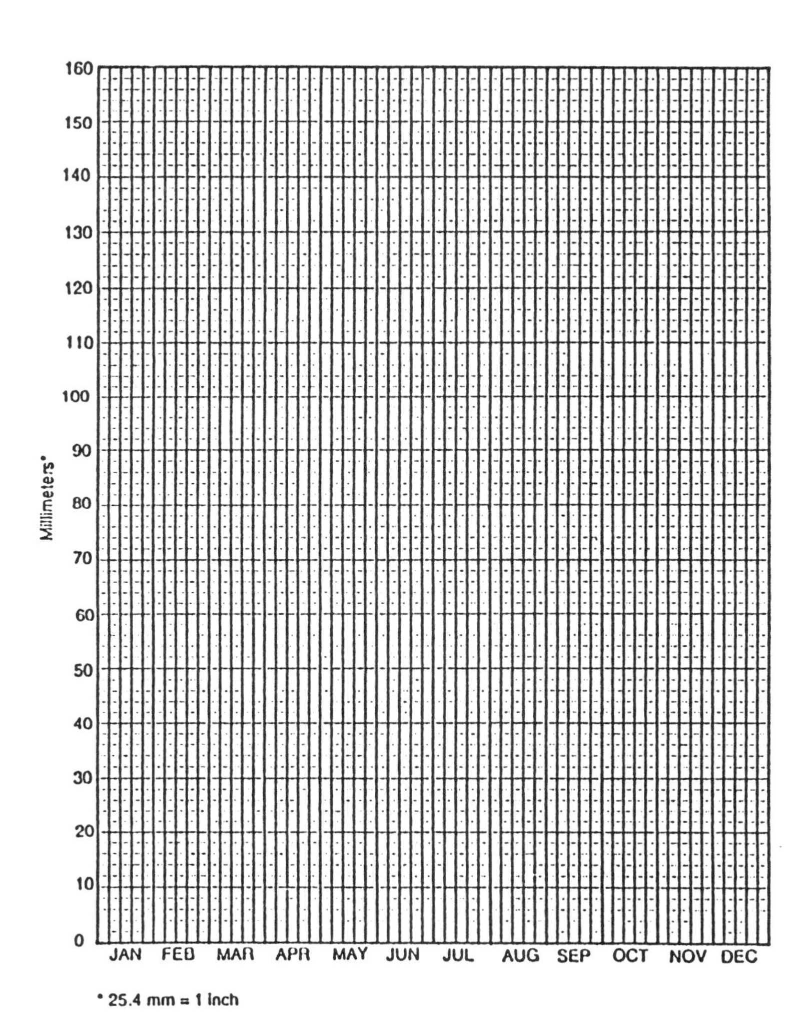

8. Compare the water balance graph you have just completed with that for Phoenix, Arizona, presented in Figure 6-9b, p. 194. Any differences or similarities between Phoenix and Kingsport?

9. The following illustration is derived from the right-half of Figure 6-12 (p. 196-97). Complete the labeling and descriptions to identify the various aspects of the groundwate environment. Use coloration to highlight the groundwater features in the illustration.

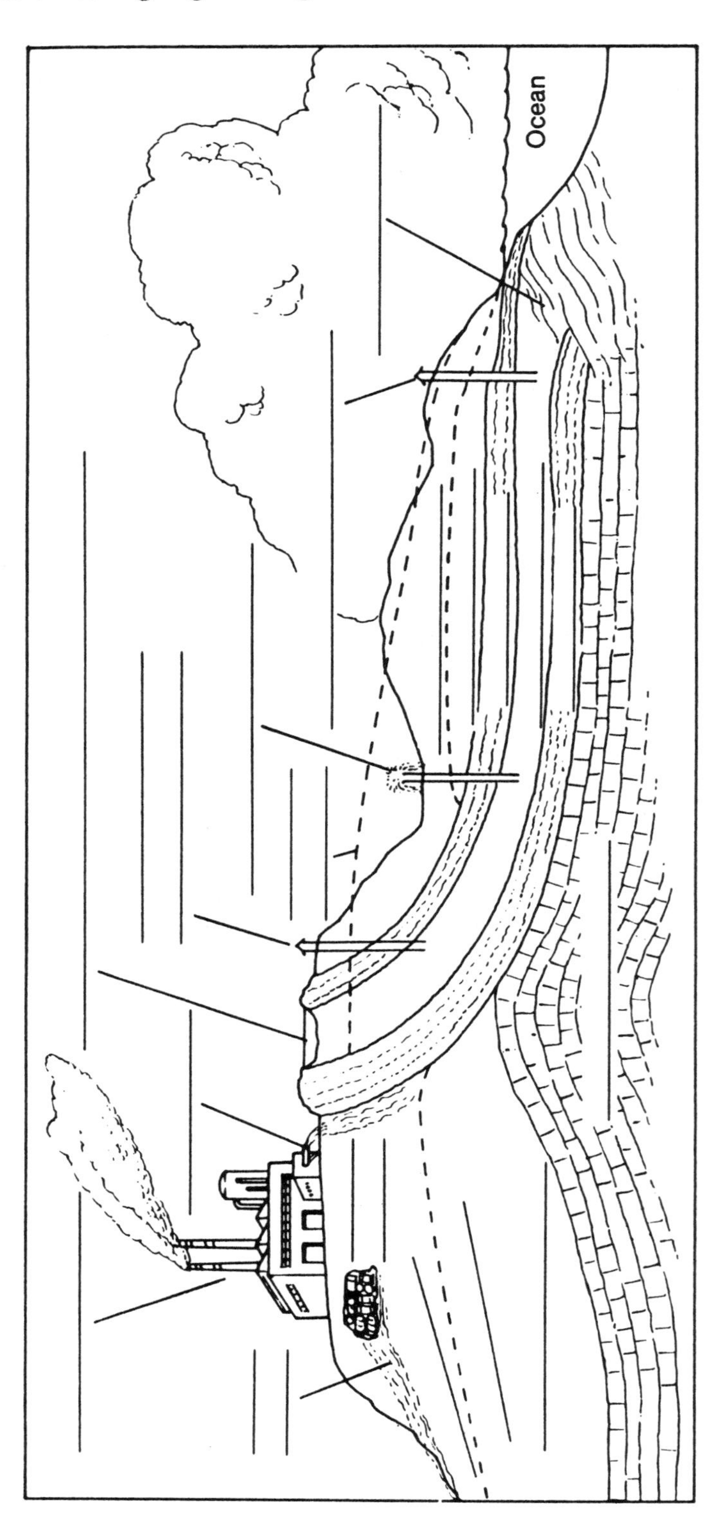

10. Figure 6-14, p. 202, portrays the distribution of runoff in the United States and Canada. Given your analysis of the PRECIP and POTET maps under item #4, briefly relate your comparison to the pattern of runoff depicted on this map.

11. Analyze the evolving status of the Ogallala Aquifer as presented in FYI Report 6-1, pp. 199-200). Then assume each of two points of view (“business as usual” and “strategies to halt the loss”) and given your analysis respond to each point of view.

a) Analysis:

b) “Business as usual”—no change in practices:

c) “Strategies to halt the loss”—change in practices:

12. Analyze the evolving “Acid Deposition: A Blight on the Landscape” as presented in FYI Report 6-2, pp. 205-06). What is acid deposition? Then assume each of two points of view “business as usual” and “strategies to halt the damage”) and given your analysis respond to each point of view.

a) Analysis:

b) “Business as usual”—no change in practices:

(c) "Strategies to halt the damage"—change in practices:____

____

____

____

13. Identify the sample of 21 oceans and seas indicated by number in Figure 6-18, p. 208. Write the correct name in the spaces provided below.

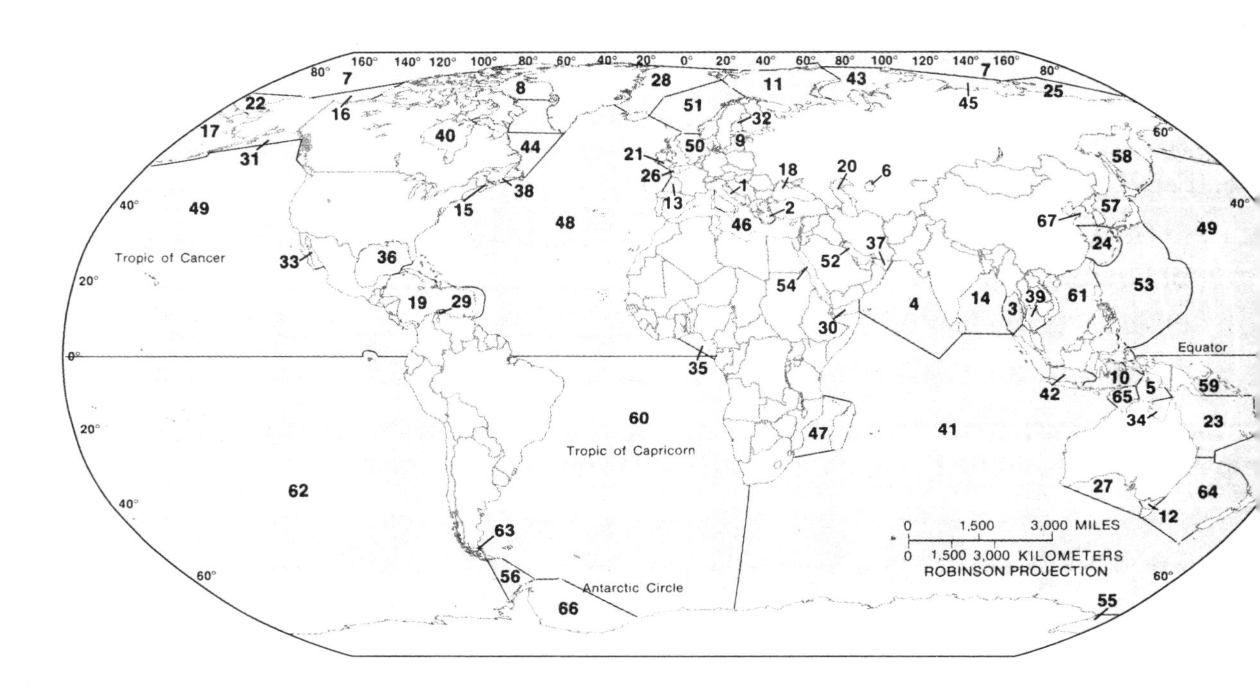

1.	36.	53.
9.	37.	54.
11.	43.	57.
14.	44.	59.
23.	46.	60.
24.	48.	64.
31.	50.	65.

3.	46.	60.
4.	48.	64.
1.	50.	65.

4. *Name* the largest ocean in terms of area by percentage of total ocean area. *Record* hat area in km^2 and volume in km^3. What is the mean depth of the ocean you have named and the deepest point in this ocean?

____

____.

5. In terms of salinity, specifically characterize the following in parts per thousand (‰):

Subtropical oceans_______________________

Equatorial oceans_______________________

Brine_______________________

Brackish_______________________

Gulf of Bothnia_______________________

Sargasso Sea_______________________

Sample Self-test

(Answers appear at the end of the study guide.)

1. Most of the precipitation and evaporation on Earth takes place over the
   a) land masses
   b) oceans and seas
   c) poles of the planet
   d) subtropical latitudes

2. Relative to the hydrologic cycle which of the following is incorrect?
   a) the bulk of the precipitation occurs over the ocean
   b) over 50 units of moisture are involved in advective flows in the hydrologic cycle model illustrated in the text
   c) 22 percent of Earth's precipitation falls over the land
   d) 78 percent of all precipitation falls on the oceans

3. Potential evapotranspiration refers to
   a) the moisture supply
   b) the amount of unmet water demand in an environment
   c) the amount of water that would evaporate or transpire if it were available
   d) the amount of water that only plants use

5. Actual evapotranspiration is determined by
    a) PRECIP – DEFIC
    b) PRECIP – SURPL
    c) ACTET – DEFIC
    d) POTET – DEFIC

6. Water detention on the surface is a form of
    a) deficit
    b) groundwater
    c) surplus
    d) soil moisture storage

7. Soil moisture that plants are capable of accessing and utilizing is called
    a) wilting point water
    b) gravitational water
    c) available water
    d) hygroscopic water

8. The largest potential source of freshwater accessible in North America is
    a) groundwater
    b) ice sheets and glaciers
    c) stream discharge
    d) potential evapotranspiration

9. A water-bearing rock strata is called
    a) soil moisture storage
    b) an aquiclude
    c) a zone of aeration
    d) an aquifer

10. The upper limit of groundwater that is available for utilization at the surface is called
    a) capillary water
    b) an aquiclude
    c) the cone of depression
    d) the water table

11. Groundwater
    a) is seemingly unlimited when compared with surface supplies
    b) when polluted, is actually easier to clean up than is surface water
    c) should be considered separately from surface supplies
    d) is reduced by the mining of water

12. After water itself (hydrogen and oxygen), what are the two primary elements that occur in se water?
    a) chlorine and sodium
    b) potassium and chlorine
    c) hydrogen and sodium
    d) oxygen and chlorine

13. The salinity of the oceans reached a concentration roughly similar to that of today about one billion years ago.
  a) true
  b) false

14. According to the text the Gulf of Bothnia is correctly referred to as brackish.
  a) true
  b) false

15. Deficit (DEFIC) is the moisture demand in the water balance.
  a) true
  b) false

16. Precipitation specifically refers to rain, sleet, snow, and hail.
  a) true
  b) false

17. Moisture entering a soil body is referred to as soil moisture utilization.
  a) true
  b) false

18. The difference between field capacity and wilting point is called capillary water, almost all of which is available for extraction by plants and evaporation.
  a) true
  b) false

19. Permeability refers to the movement of water through soil or porous rock.
  a) true
  b) false

20. The rate of flow of a river is simply called its discharge.
  a) true
  b) false

21. Relative to water usage in the United States (lower 48 states), we currently are withdrawing 10% of the available surplus.
  a) true b) false

Name:________________________________ Class Section:________________

Date:__________________ Score/Grade:______________

Earth's Climates

7

Chapter Overview

Chapter 7 serves as a synthesis of content from Chapters 2 through 6—Parts One and Two of the text. Earth experiences an almost infinite variety of weather. Even the same location may go through periods of changing weather. This variability, when considered along with the average conditions at a place over time, constitutes climate. Climates are so diverse that no two places on Earth's surface experience exactly the same climatic conditions, although general similarities permit grouping and classification. The chapter also looks at how Earth's temperature system appears to be in a state of dynamic change as concerns about global warming and potential episodes of global cooling are discussed. This chapter presents an overview of the climatic effects of these temperature changes.

Learning Objectives

The following learning objectives help guide your reading and comprehension efforts. The operative word is in *italics*. Read and work with these carefully and note that exercise #1 asks you about five of these objectives. After reading the chapter and using this workbook, you should be able to:

1. *Define* climate and climatology and *discern* the difference between climate and weather.
2. *Review* the role of temperature and precipitation, pressure, and air mass patterns as basic inputs toward establishing climatic regions.
3. *Analyze* world precipitation patterns and *discern* essential causes for these patterns based on Chapters 5 and 6.
4. *Compare* and *contrast* world temperature maps (Figures 3-15 and 3-17, pp. 94, 96) and world precipitation maps (Figure 6-4, p. 190, for the United States and Canada and Figure 7-2, p. 218, for the world).
5. *Analyze* the El Niño/Southern Oscillation phenomena and *relate* the principal occurrences in this century, in particular, the 1982-83 event.
6. *Explain* classification and *describe* climatic classification in particular.
7. *Relate* the story of Wladimir Köppen and *review* his development of a climate classification system.
8. *Describe* the classification criteria used by Köppen in determining Earth's climates.
9. *Relate* the determination of the B climates to moisture efficiency and *explain* the role of seasonal precipitation distribution.

10. *Differentiate* between A, C, D, and E climate classifications using temperature criteria and *list* the individual climate types derived.
11. *Construct* a climograph and *plot* temperature and precipitation data for several weather stations (the necessary data are in Appendix A).
12. *Correlate* the patterns of world climates and precipitation on the maps and *relate* them to the map of ten major biomes in Figure 16-22, p. 506.
13. *Describe* the distribution of Earth's equatorial and tropical rain forests.
14. *Explain* the occurrence of a marine west coast climate in the Appalachians.
15. *Analyze* the association of Mediterranean dry-summer climates and west coast coastal fog in the midlatitudes.
16. *Portray* the climate of Verkhoyansk, Russia, using information in Figure 7-15, p. 233, and Figure 3-16, p. 95.
17. *Identify* the principal desert regions of the world as defined in the Köppen system.
18. *Outline* future climate patterns and *explain* the operative causes and processes that are suspected.
19. *Review* potential consequences of climatic warming in the midlatitudes and high latitudes.
20. *List* specific solutions to slow climate change as detailed in the text and in your own opinion.
21. *Describe* the structure of a general circulation model (GCM). *List* the four GCM models in operation in the United States and four in other countries.
22. *Outline* the basic anthropogenic factors involved in global warming and *include* an explanation of present indicators and future consequences.
23. *Describe* the concentration of CO_2 in the lower atmosphere from 1825 to the present.
24. *Analyze* the nuclear winter hypothesis and *explain* some potential consequences.

Outline Headings and Glossary Review

These are the first- second-, and third-order headings that divide this chapter. The key terms and concepts that appear **boldface** in the text are listed under their appropriate heading in bold italics; these highlighted terms appear in the text glossary. A check-off box is placed next to each key term so you can mark your progress through the chapter as you define these in your reading notes or prepare note cards.

Climate System Components

❑ ***climate***
❑ ***climatology***
❑ ***climatic regions***

Temperature and Precipitation

Classification of Climatic Regions

❑ ***classification***
❑ ***genetic classification***
❑ ***empirical classification***

The Köppen Classification System
Classification Criteria
❑ ***Köppen-Geiger climate classification***
Köppen's Climatic Designations
Global Climate Patterns
❑ ***climographs***
Tropical A Climates
Tropical Rain Forest Climates (Af)
Tropical Monsoon Climates (Am)
Tropical Savanna Climates (Aw)
Mesothermal C Climates
Humid Subtropical Hot Summer Climates (Cfa, Cwa)
Marine West Coast Climates (Cfb, Cfc)
Mediterranean Dry Summer Climates (Csa, Csb)
Microthermal D Climates
Humid Continental Hot Summer Climates (Dfa, Dwa)
Humid Continental Mild Summer Climates (Dfb, Dwb)
Subarctic Climates (Dfc, Dwc, Dwd)
Polar E Climates
Dry Arid and Semiarid B Climates
Hot Low-Latitude Desert Climates (BWh)

Cold Midlatitude Desert Climates (BWk)
Hot Low-Latitude Steppe Climates (BSh)
Cold Midlatitude Steppe Climates (BSk)

Future Climate Change

Developing Climate Models
❑ ***general circulation models (GCMs)***

Global Warming
❑ ***paleoclimatolgy***

Carbon Dioxide and Global Warming

Methane and Global Warming
❑ ***methane***

Chlorofluorocarbons (CFCs) and Global Warming

Warming Indications and the Future

Consequences of Climatic Warming

Effects on World Food Supply and the Biosphere

Melting Glaciers and Ice Sheets

Solutions

Global Cooling
❑ ***nuclear winter hypothesis***

SUMMARY

Learning Activities and Critical Thinking

1. Select any five learning objectives from the list presented at the beginning of this chapter. Place the number selected in the space provided (no need to rewrite each objective). Using the following questions as guidelines only, briefly discuss your treatment of the objective.

- What did you know about the objective before you began?
- What was your plan to complete the objective?
- Which information source did you use in your learning (text, or other)?
- Were you able to complete the action stated in the objective? What did you learn?
- Are there any aspects of the objective about which you want to know more?

a)____:____

____

____

____

____

____.

b)____:____

____

____

____

____

____.

c)____ :____

____

____

____

____

____.

d)____ :____

____

____

____

____

____.

e)____ :____

____

____

____

____

____.

2. Distinguish between *weather* and *climate* by defining each.

(a) weather:____

(b) climate:____

3. What is the difference between an *empirical* and a *genetic classifications*; use climatic factors as an example (pp. 218-19)?

____

____

____

____.

4. What is the appropriate Köppen climate designation for your present location? If this is different from your home town, also list the climate type for your home. See the "Geography I.D." page in the introduction to this study guide.

____.

5. Given your answer in #2 (your Köppen classification) and using the following schematic of temperature and precipitation interactions from Figure 7-3, p. 219, mark your present location (approximately) with an "**X**" on the temperature and precipitation schematic below. Place the name of this location next to the mark. Next, mark the location (approximately) most characteristic of your birthplace with an "**O**" and record the name of this place next to the mark.

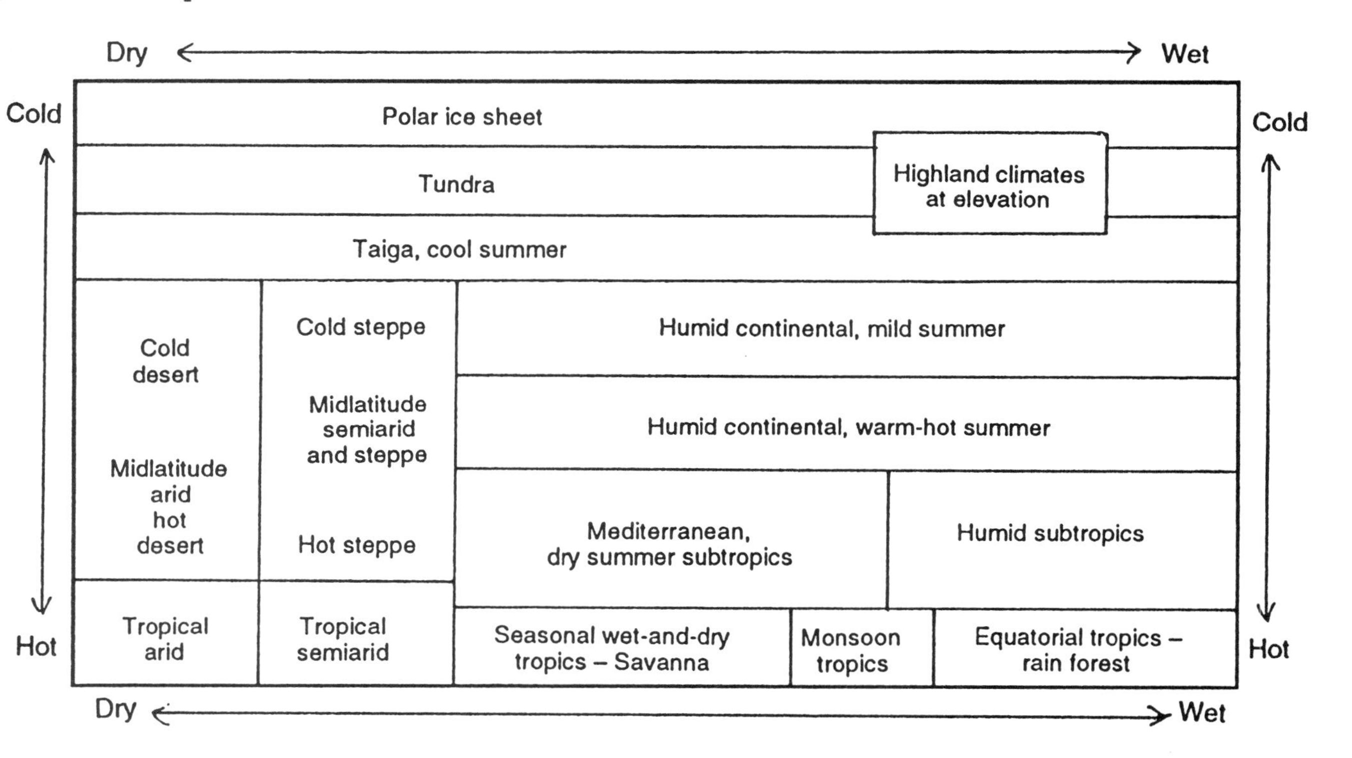

6. Which Köppen classification designation is based on the amount and seasonal distribution of precipitation?_____. What three precipitation regimes (annual distribution) are described in Figure 7-4, pp. 220-21, for the B climate classification?

a)____

b)____

c)____.

7. Below are average annual data for **a)** Alma-Ata, Kazakhstan, and **b)** Maracaibo, Venezuela. Determine the Köppen classification for each city using the key information in Figure 7-4, pp. 220-21, in the text. You may disregard the POTET (potential evapotranspiration) data.

a) Alma-Ata______________________________ **b)** Maracaibo________________________________

Alma-Ata, Kazakhstan: pop. 910,000, lat. 43° 15'N, long. 76° 57'E, elev. 847 m (2,779 ft).

	Jan	Feb	Mar	Apr	May	Jun	Jul	Aug	Sep	Oct	Nov	Dec	Annual
Temperature °C	-7.9	-5.1	1.6	10.8	16.0	20.4	23.3	22.3	17.4	10.0	-0.1	-5.4	8.7
(°F)	(17.8)	(22.8)	(34.9)	(51.4)	(60.8)	(68.7)	(73.9)	(72.1)	(63.3)	(50.0)	(31.8)	(22.3)	(47.7)
PRECIP cm	2.5	3.3	6.4	8.9	9.9	5.8	3.6	2.3	2.5	4.6	4.8	3.6	58.2
(in.)	(1.0)	(1.3)	(2.5)	(3.5)	(3.9)	(2.3)	(1.4)	(0.9)	(1.0)	(1.8)	(1.9)	(1.4)	(22.9)
POTET cm	0	0	0.5	5.1	9.1	12.7	15.2	13.5	8.6	4.1	0	0	68.1
(in.)	(0)	(0)	(0.2)	(2.0)	(3.6)	(5.0)	(6.0)	(5.3)	(3.4)	(1.6)	(0)	(0)	(26.8)

Maracaibo, Venezuela: pop. 900,000, lat. 10° 40'N, long. 71° 37'W, elev. 40 m (131 ft).

	Jan	Feb	Mar	Apr	May	Jun	Jul	Aug	Sep	Oct	Nov	Dec	Annual
Temperature °C	26.5	26.7	27.2	27.8	28.4	28.6	28.6	28.7	28.7	27.8	27.8	27.1	27.9
(°F)	(79.7)	(80.1)	(81.0)	(82.0)	(83.1)	(83.5)	(83.5)	(83.7)	(83.7)	(82.2)	(82.0)	(80.8)	(82.2)
PRECIP cm	0.3	0.3	0.3	2.5	7.4	4.3	2.8	4.1	4.1	9.9	2.3	0.5	38.6
(in.)	(0.1)	(0.1)	(0.1)	(1.0)	(2.9)	(1.7)	(1.1)	(1.6)	(1.6)	(3.9)	(0.9)	(0.2)	(15.2)
POTET cm	12.7	12.4	14.5	15.0	16.3	16.3	16.8	16.5	15.5	15.0	14.0	13.7	178.8
(in.)	(5.0)	(4.9)	(5.7)	(5.9)	(6.4)	(6.4)	(6.6)	(6.5)	(6.1)	(5.9)	(5.5)	(5.4)	(70.4)

8. Use the six climographs on the next 6 pages and sets of data to plot: **mean monthly temperature** (*solid-line graph*), **mean monthly precipitation** (*a bar graph*), and **potential evapotranspiration** (*a dotted line*) data for the following stations. Use the climate key in Figure 7-4, pp. 220-21, to determine each station's classification. Analyze the distribution of temperature and precipitation during the year and record the other information as requested to complete each climograph.

(a) **A** climate: Salvador (Bahia), Brazil (#1)
(b) **Cf** climate: New Orleans, Louisiana (#2)
(c) **Cs** climate: Sacramento, California (#3)
(d) **D** climate: Montreal, Quebec, Canada (#4)
(e) **BW** climate: Yuma, Arizona (#5)
(f) **BS** climate: Kemmerer, Wyoming (#6)

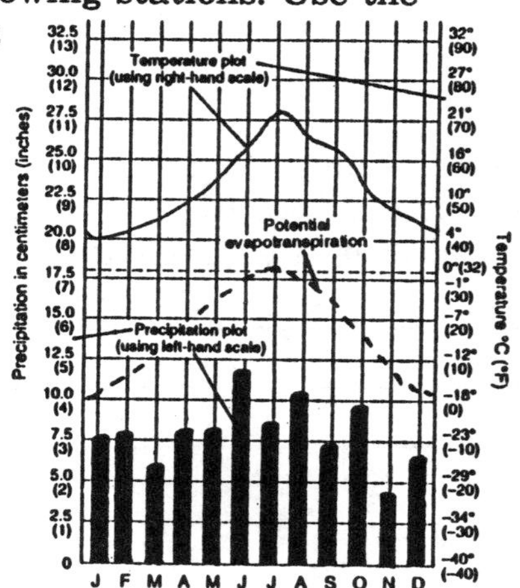

(8a.) A climate: Salvador (Bahia), Brazil (#1)

Latitude____________________________

Longitude____________________________

Elevation____________________________

Population____________________________

Total annual rainfall:____________________________

Average annual temperature:____________________________

Annual temperature range:____________________________

Distribution of temperature during the year:

____

____________________________.

Distribution of precipitation during the year:

____

____________________________.

Distribution of potential evapotranspiration during the year:____________________________

____________________________.

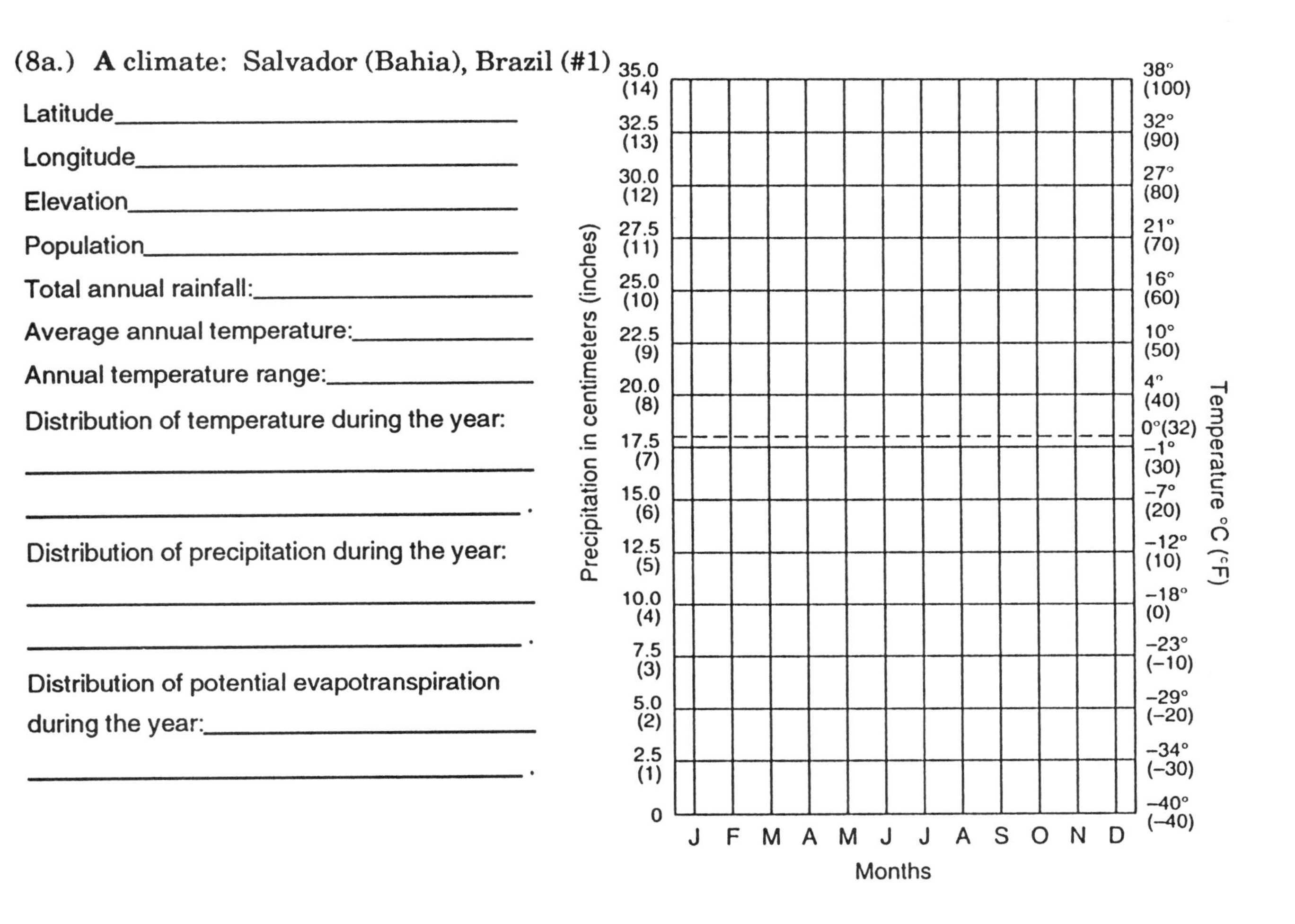

Salvador (Bahia), Brazil: pop. 1,507,000, lat. 12° 59'S, long. 38° 31'W, elev. 9 m (30 ft).

	Jan	Feb	Mar	Apr	May	Jun	Jul	Aug	Sep	Oct	Nov	Dec	Annual
perature °C (°F)	26.0 (78.8)	26.3 (79.3)	26.3 (79.3)	25.8 (78.4)	24.8 (76.6)	23.8 (74.8)	23.0 (73.4)	22.9 (73.2)	23.6 (74.5)	24.5 (76.1)	25.1 (77.2)	25.6 (78.1)	24.8 (76.6)
ECIP cm (in.)	7.4 (2.9)	7.9 (3.1)	16.3 (6.4)	29.0 (11.4)	29.7 (11.7)	19.6 (7.7)	20.6 (8.1)	11.2 (4.4)	8.4 (3.3)	9.4 (3.7)	14.2 (5.6)	9.9 (3.9)	183.6 (72.3)
TET cm (in.)	13.7 (5.4)	12.4 (4.9)	13.5 (5.3)	11.9 (4.7)	10.7 (4.2)	9.1 (3.6)	8.6 (3.4)	8.6 (3.4)	9.4 (3.7)	11.2 (4.4)	11.9 (4.7)	13.2 (5.2)	134.4 (52.9)

Köppen climate classification symbol:________; name:____________________________

____________________; explanation for this determination:____________________________

____

____.

(8b.) C climate: New Orleans, Louisiana (#2)

Latitude_______________

Longitude_______________

Elevation_______________

Population_______________

Total annual rainfall:_______________

Average annual temperature:_______________

Annual temperature range:_______________

Distribution of temperature during the year:

____

_______________.

Distribution of precipitation during the year:

____

_______________.

Distribution of potential evapotranspiration during the year:_______________

_______________.

New Orleans, Louisana: pop. 557,000, lat. 29° 57'N, long. 90 ° 04'W, elev. 3 m (9 ft).

	Jan	Feb	Mar	Apr	May	Jun	Jul	Aug	Sep	Oct	Nov	Dec	An
Temperature °C	13.3	14.4	17.2	21.1	24.4	27.8	28.3	28.3	26.7	22.8	16.7	13.9	21
(°F)	(56.0)	(58.0)	(63.0)	(70.0)	(76.0)	(82.0)	(83.0)	(83.0)	(80.0)	(73.0)	(62.0)	(57.0)	(70
PRECIP cm	12.2	10.7	16.8	13.7	13.7	14.2	18.0	16.3	14.7	9.4	10.2	11.7	161
(in.)	(4.8)	(4.2)	(6.6)	(5.4)	(5.4)	(5.6)	(7.1)	(6.4)	(5.8)	(3.7)	(4.0)	(4.6)	(63
POTET cm	2.2	2.6	4.9	8.4	12.7	16.8	18.0	17.1	13.9	8.8	4.0	2.4	111
(in.)	(0.9)	(1.0)	(1.9)	(3.3)	(5.0)	(6.6)	(7.1)	(6.7)	(5.5)	(3.5)	(1.6)	(0.9)	(44

Köppen climate classification symbol:________; name:_______________

_______________; explanation for this determination:_______________

____

_______________.

(8c.) **C climate:** Sacramento, California (#3)

Latitude________________________________

Longitude________________________________

Elevation________________________________

Population________________________________

Total annual rainfall:________________________

Average annual temperature:__________________

Annual temperature range:____________________

Distribution of temperature during the year:

____

____.

Distribution of precipitation during the year:

____

____.

Distribution of potential evapotranspiration during the year:_____________________________

____.

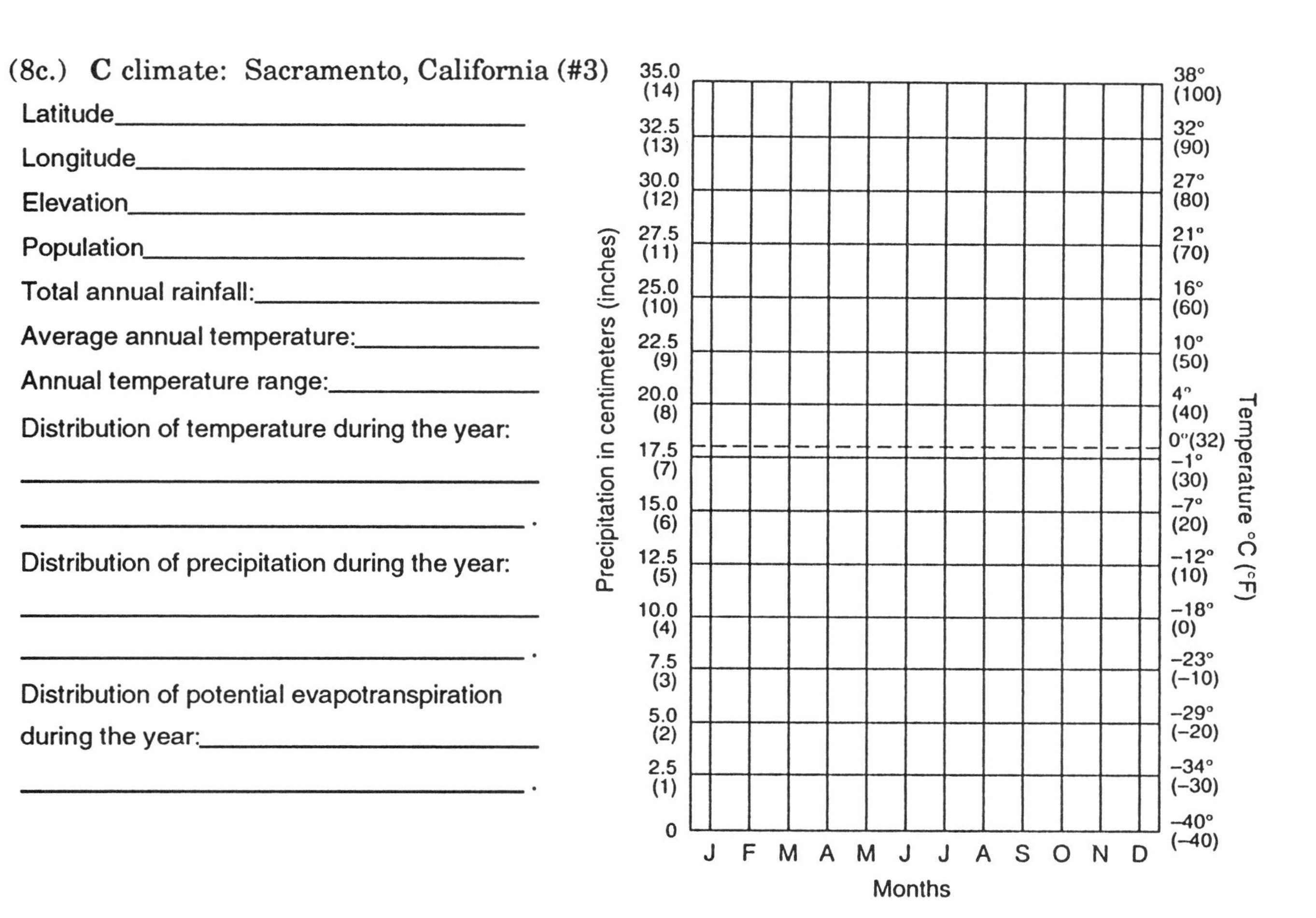

Sacramento, California: pop. 330,000, lat. 38° 35'N, long. 121° 21'W, elev. 11 m (36 ft).

	Jan	Feb	Mar	Apr	May	Jun	Jul	Aug	Sep	Oct	Nov	Dec	Annual
erature°C (°F)	8.4 (47.1)	11.2 (52.2)	12.9 (55.3)	15.6 (60.1)	19.1 (66.3)	22.3 (72.2)	24.8 (76.6)	24.2 (75.6)	22.7 (72.9)	18.5 (65.3)	12.6 (54.7)	8.6 (47.5)	16.8 (62.2)
CIP cm (in.)	10.7 (4.2)	7.4 (2.9)	5.6 (2.2)	3.6 (1.4)	1.0 (0.4)	0.3 (0.1)	0.3 (0.1)	0.3 (0.1)	0.8 (0.3)	2.3 (0.9)	5.8 (2.3)	7.6 (3.0)	45.5 (17.9)
ET cm (in.)	1.5 (0.6)	2.3 (0.9)	4.0 (1.6)	6.0 (2.4)	8.5 (3.3)	11.5 (4.5)	14.0 (5.5)	12.7 (5.0)	9.9 (3.9)	6.6 (2.6)	3.0 (1.2)	1.5 (0.6)	81.5 (32.1)

Köppen climate classification symbol:________; name:____

____________________; explanation for this determination:______________________________________

____

____.

(8d.) **D** climate: Montreal, Quebec, Canada (#4)

Latitude________________________________

Longitude________________________________

Elevation________________________________

Population________________________________

Total annual rainfall:________________________

Average annual temperature:________________

Annual temperature range:__________________

Distribution of temperature during the year:

____

____.

Distribution of precipitation during the year:

____

____.

Distribution of potential evapotranspiration

during the year:____________________________

____.

Precipitation in centimeters (inches)

35.0 (14) | 32.5 (13) | 30.0 (12) | 27.5 (11) | 25.0 (10) | 22.5 (9) | 20.0 (8) | 17.5 (7) | 15.0 (6) | 12.5 (5) | 10.0 (4) | 7.5 (3) | 5.0 (2) | 2.5 (1) | 0

38° (100) | 32° (90) | 27° (80) | 21° (70) | 16° (60) | 10° (50) | 4° (40) | 0°(32 | −1° (30) | −7° (20) | −12° (10) | −18° (0) | −23° (−10) | −29° (−20) | −34° (−30) | −40° (−40)

J F M A M J J A S O N D

Months

Montreal, Quebec, Canada: pop. 2,818,000, lat. 45° 30'N, long 73° 35'W, elev. 57 m (187 ft).

	Jan	Feb	Mar	Apr	May	Jun	Jul	Aug	Sep	Oct	Nov	Dec	A
Temperature °C	-10.0	-9.4	-3.3	5.6	13.3	18.3	22.1	19.4	15.0	8.3	0.6	-6.7	
(°F)	(14.0)	(15.1)	(26.1)	(42.1)	(55.9)	(64.9)	(70.0)	(66.9)	(59.0)	(46.9)	(33.1)	(19.9)	(4
PRECIP cm	9.6	7.7	8.8	6.6	8.0	8.7	9.5	8.8	9.3	8.7	9.0	9.1	10
(in.)	(3.8)	(3.0)	(3.5)	(2.6)	(3.1)	(3.4)	(3.7)	(3.5)	(3.7)	(3.4)	(3.5)	(3.6)	(4
POTET cm	0	0	0	2.7	8.1	11.9	13.9	12.1	7.7	3.7	0.2	0	6
(in.)	(0)	(0)	(0)	(1.1)	(3.2)	(4.7)	(5.5)	(4.8)	(3.0)	(1.5)	(0.1)	(0)	(2

Köppen climate classification symbol:________; name:________________________________

__________________; explanation for this determination:______________________________

____

____.

(8e.) **B climate: Yuma, Arizona (#5)**

Latitude________________________________

Longitude_______________________________

Elevation_______________________________

Population______________________________

Total annual rainfall:____________________

Average annual temperature:_______________

Annual temperature range:_________________

Distribution of temperature during the year:

____

____.

Distribution of precipitation during the year:

____

____.

Distribution of potential evapotranspiration during the year:__________________

____.

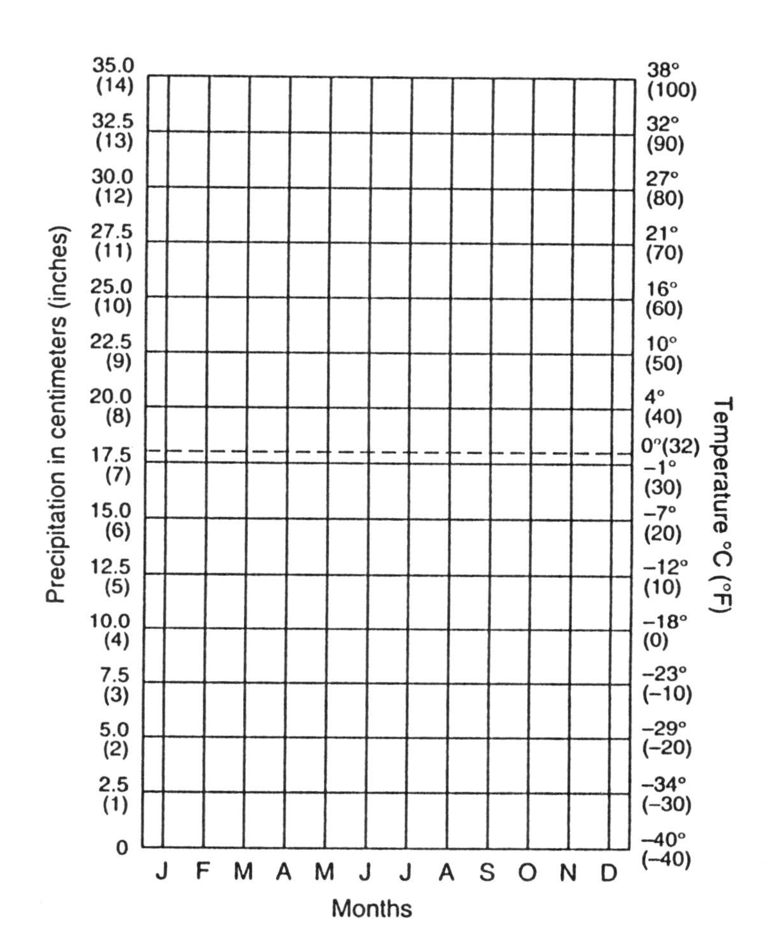

Yuma, Arizona: pop. 42,000, lat. 32° 40′N, long. 114° 36′W, elev. 61 m (200 ft).

	Jan	Feb	Mar	Apr	May	Jun	Jul	Aug	Sep	Oct	Nov	Dec	Annual
erature°C (°F)	12.7 (54.9)	14.8 (58.6)	17.9 (64.2)	21.5 (70.7)	25.1 (77.2)	29.6 (85.3)	33.2 (91.8)	32.8 (91.0)	29.7 (85.5)	23.3 (73.9)	17.1 (62.8)	13.2 (55.8)	22.6 (72.7)
CIP cm (in.)	1.1 (0.4)	1.1 (0.4)	0.8 (0.3)	0.3 (0.1)	0.1 (0)	0 (0)	0.5 (0.2)	1.4 (0.6)	1.0 (0.4)	0.8 (0.3)	0.6 (0.2)	1.2 (0.5)	8.9 (3.5)
ET cm (in.)	1.3 (0.5)	2.3 (0.9)	4.6 (1.8)	8.2 (3.2)	13.6 (5.4)	18.9 (7.4)	21.1 (8.3)	20.0 (7.9)	16.4 (6.5)	9.0 (3.5)	3.4 (1.3)	1.5 (0.6)	110.3 (43.4)

Köppen climate classification symbol:________; name:________________________________

____________________; explanation for this determination:______________________________

____

____.

(8f.) **B** climate: Kemmerer, Wyoming (#6)

Latitude________________________________

Longitude________________________________

Elevation________________________________

Population________________________________

Total annual rainfall:________________________________

Average annual temperature:________________________________

Annual temperature range:________________________________

Distribution of temperature during the year:

____

____.

Distribution of precipitation during the year:

____

____.

Distribution of potential evapotranspiration during the year:________________________________

____.

Precipitation in centimeters (inches)

35.0 (14)
32.5 (13)
30.0 (12)
27.5 (11)
25.0 (10)
22.5 (9)
20.0 (8)
17.5 (7)
15.0 (6)
12.5 (5)
10.0 (4)
7.5 (3)
5.0 (2)
2.5 (1)
0

Temperature °C (°F)

38° (100)
32° (90)
27° (80)
21° (70)
16° (60)
10° (50)
4° (40)
0°(32)
−1° (30)
−7° (20)
−12° (10)
−18° (0)
−23° (−10)
−29° (−20)
−34° (−30)
−40° (−40)

J F M A M J J A S O N D

Months

Kemmerer, Wyoming: pop. 3300, lat. 41° 48'N, long. 110° 32'W, elev. 2120 m (6954 ft).

	Jan	Feb	Mar	Apr	May	Jun	Jul	Aug	Sep	Oct	Nov	Dec	An
Temperature °C	-8.3	-6.6	-2.8	3.9	9.2	13.3	17.2	16.1	11.5	5.7	-1.9	-5.5	4
(°F)	(17.0)	(20.1)	(27.0)	(39.0)	(48.6)	(55.9)	(63.0)	(60.9)	(52.7)	(42.3)	(28.5)	(22.0)	(39
PRECIP cm	1.8	1.7	1.9	2.1	2.6	2.8	1.9	2.1	1.6	2.1	1.7	1.7	24
(in.)	(0.7)	(0.7)	(0.7)	(0.8)	(1.0)	(1.1)	(0.7)	(0.8)	(0.6)	(0.8)	(0.7)	(0.7)	(9
POTET cm	0	0	0	2.7	6.4	9.1	11.9	10.4	6.6	3.4	0	0	50
(in.)	(0)	(0)	(0)	(1.1)	(2.5)	(3.6)	(4.7)	(4.1)	(2.6)	(1.3)	(0)	(0)	(19

Köppen climate classification symbol:________; name:____

__________________; explanation for this determination:________________________________

____

____.

9. Briefly describe the polar environment. What are the Köppen classifications for these regions?

____

____

____.

10. Check the listings in the Index of the text for the following and briefly define and describe each:

(a) Arctic region:____

____.

(b) Antarctic region:____

____.

(c) Periglacial landscapes:____

____.

11. What is a "GCM"? List the eight operating systems.

____

____

____

____.

12. When were the nine warmest years in instrumental history?______________________.
In this context, examine the temperature graph in Figure 7-23, p. 242, and the GISS–GCM projections shown in Figure 7-24, p. 243.

13. Describe the trend for atmospheric carbon dioxide (CO_2) between the years 1825 and 2025. ____

____

____.

14. Characterize countries and regions as to their production of excessive CO_2 (Figure 7-22, p. 240) for each of the following years:

a) 1980:____.

b) 2025 (forecast):____.

c) What sectors will show the greatest increase in production over the 45-year period?

____.

15. Identify and describe at least <u>three</u> potential consequences of a possible global warming.

a)____

____

____

b)____

____

____

c)____

____

____

16. In terms of global cooling, what was the impact of the Mount Pinatubo eruption in June 1991 on temperatures and radiation balances in the atmosphere in the two years following the blast? (In addition to Chapter 10, p. 244 and Figure 7-23, p. 242, also refer t text pages 7, 53, 78, 109, 308, and 538.)

____

____

____

____

Sample Self-test

(Answers appear at the end of the study guide.)

1. An area that contains characteristic weather patterns is called a/an
   a) climatology
   b) El Ninõ, or ENSO
   c) weather phenomenon
   d) climatic region

2. An empirical classification is partially based on
   a) the interaction of air masses
   b) the origin or genesis of the climate
   c) mean annual temperature and precipitation
   d) causative factors

3. Relative to tropical A climates, which of the following is true?
   a) all months average below 18°C (64.4°F)
   b) annual POTET exceeds PRECIP
   c) strong seasonality prevails
   d) all months average warmer than 18°C (64.4°F)

4. Relative to the Csa classification, which of the following is false?
   a) summers are hot
   b) 70% of the PRECIP occurs in the winter months
   c) it is also called the Mediterranean dry-summer climate
   d) its warmest summer month averages below 22°C (71.6°F)

5. The coldest climate on Earth, outside of the polar regions is the
   a) Dwc
   b) Dfc
   c) EF
   d) Dwd

6. Relative to an EF climate, which of the following is correct?
   a) the annual temperature range is less than 17C° (30F°)
   b) it is generally called the tundra
   c) its warmest month is below 0°C (32°F)
   d) its warmest month is above 0°C (32°F)

7. If PRECIP is more than 1/2 POTET but not equal to it, the climate is considered a/an
   a) Af
   b) BW
   c) Bk
   d) BS

8. The most extensive climates, occupying the largest percentage of Earth's surface, are the
   a) A climates
   b) C climates
   c) D climates
   d) E climates

9. Which of the following is a typical Dwd climate?
   a) Churchill, Manitoba
   b) Lisbon, Portugal
   c) Dalian, China
   d) Verkhoyansk, Siberia, Russia

10. According to the IPCC assessment, p. 241, temperatures by the year A.D. 2029 will be
    a) 2.5C° higher
    b) unrelated to the behavior of society
    c) 1.5C° lower
    d) 4.5C° higher

11. Relative to future temperatures,
   a) humans can not influence long-term temperature trends
   b) short-term changes appear to be out of our reach to influence
   c) a cooperative global network of weather monitoring among nations has yet to be established
   d) human society appears to be causing short-term changes in global temperatures and temperature patterns
12. The warmest years in the history of weather instruments were
   a) recorded during the period 1910 to 1921
   b) between 1980 and 1993 with nine of the warmest years
   c) in the 1950s
   d) not determined since temperatures are not exhibiting any trend at this time

13. According to the quote (p, 239) from Richard Houghton and George Woodwell in *Scientific American*, climate zones are not shifting nor is sea level rising at this time.
   a) true
   b) false

14. The Intergovernmental Panel on Climate Change (IPCC) has reached virtual unanimity among greenhouse experts that a climatic warming is occurring; uncertainty exists as to the severity and exact timing of consequences.
   a) true
   b) false

15. The Köppen climatic classification system is an example of a genetic classification.
   a) true
   b) false

16. The only Köppen classification that is based on moisture as well as temperature includes the H climates.
   a) true
   b) false

17. The C climates make up the second-largest percentage of Earth's surface (land and water areas) and 55% of Earth's resident population.
   a) true
   b) false

18. Subarctic climates include regions of the highest degree of continentality on Earth.
   a) true
   b) false

19. POTET exceeds PRECIP in all parts of the B climates with no exceptions.
   a) true   b) false

20. Most of the Sahara is characterized by the BSk classification.
   a) true   b) false

PART THREE:
Earth's Changing Landscapes

Overview–Part Three

Earth is a dynamic planet whose surface is actively shaped by physical agents of change. Part Three is organized around two broad systems of these agents—the endogenic, or internal, system, and the exogenic, or external, system. The endogenic system (Chapters 8 and 9) encompasses processes that produce flows of heat and material from deep below the crust and are powered by radioactive decay—this is the solid realm of Earth.

The exogenic system (Chapters 10 to 14) includes processes that set air, water, and ice into motion and are powered by solar energy—this is the fluid realm of Earth's environment. Thus, Earth's surface is the interface between two systems, one that builds the landscape and one that reduces it. Both are subjects of Part Three.

Name:______________________ Class Section:______________

Date:______________ Score/Grade:__________

The Dynamic Planet

Chapter Overview

The Twentieth Century is a time of great discovery about Earth's internal structure and dynamic crust, yet much remains undiscovered. This is a time of revolution in our understanding of how the present arrangement of continents and oceans evolved. A new era of Earth-systems science is emerging, effectively combining various disciplines within the study of physical geography. The geographic essence of geology, geophysics, paleontology, seismology, and geomorphology are all integrated by geographers to produce an overall picture of Earth's surface environment.

Learning Objectives

The following learning objectives help guide your reading and comprehension efforts. The operative word is in *italics*. Read and work with these carefully and note that exercise #1 asks you about five of these objectives. After reading the chapter and using this workbook, you should be able to:

1. *Distinguish* between the endogenic and exogenic systems, the driving force for each, and the agents of energy and mass transfer.
2. *Outline* the geologic time scale and *identify* relative and absolute dating.
3. *Describe* several important events in Earth's life history.
4. *Define* and *contrast* uniformitarianism and catastrophism.
5. *Diagram* Earth in cross section and *identify* the distinct internal layers.
6. *Define* Earth's core, outer core, mantle, asthenosphere, and crust.
7. *Analyze* the generation of Earth's magnetic field.
8. *Explain* the magnetic reversal phenomena and *describe* how it is used to interpret crustal movement.
9. *Describe* the interaction between the asthenosphere and crust that drives plate tectonics.
10. *Distinguish* between continental crust and oceanic crust.
11. *Relate* the principle of isostasy to crustal depression and uplift.
12. *Illustrate* the geologic cycle schematically and *relate* the rock cycle specifically to endogenic and exogenic processes.
13. *Define* igneous, metamorphic, and sedimentary processes.
14. *Identify* intrusive and extrusive igneous formations.
15. *Construct* the scientific history of continental drift in this century and *describe* the principal discoveries that led to the theory of plate tectonics.
16. *Describe* Pangaea and the breakup of Pangaea.
17. *Explain* sea-floor spreading and subduction.
18. *Outline* plate tectonics as an inclusive conceptual model.
19. *Distinguish* between the three types of plate interactions and *relate* specific geographic examples of each.
20. *Portray* the pattern of Earth's major plates and *relate* this to the occurrence of earthquakes and volcanic activity.
21. *Locate* principal hot spots on Earth and *describe* specifically the formation of the Hawaiian-Emperor Island chain.

Outline Headings and Glossary Review

These are the first- second-, and third-order headings that divide this chapter. The key terms and concepts that appear **boldface** in the text are listed under their appropriate heading in bold italics; these highlighted terms appear in the text glossary. A check-off box is placed next to each key term so you can mark your progress through the chapter as you define these in your reading notes or prepare note cards.

The Pace of Change

- ❑ ***geologic time scale***
- ❑ ***uniformitarianism***

Earth's Structure and Internal Energy

Earth in Cross Section
- ❑ ***seismic waves***

Earth's Core
- ❑ ***core***

Earth's Magnetism
- ❑ ***magnetic reversal***

Earth's Mantle
- ❑ ***mantle***

Lithosphere and Crust
- ❑ ***crust***
- ❑ ***Mohorovicic discontinuity***
- ❑ ***Moho***
- ❑ ***granite***
- ❑ ***basalt***
- ❑ ***isostasy***

Geologic Cycle

- ❑ ***geologic cycle***

Rock Cycle

- ❑ ***mineral***
- ❑ ***rock***
- ❑ ***rock cycle***

Igneous Processes

- ❑ ***igneous rocks***
- ❑ ***magma***
- ❑ ***lava***
- ❑ ***pluton***
- ❑ ***batholith***

Sedimentary Processes

- ❑ ***stratigraphy***
- ❑ ***sedimentary rocks***
- ❑ ***lithification***
- ❑ ***limestone***
- ❑ ***coal***
- ❑ ***evaporites***

Metamorphic Processes

- ❑ ***metamorphic rock***
- ❑ ***tectonic processes***

Continental Drift and Plate Tectonics

A Brief History

- ❑ ***Pangaea***

Sea-Floor Spreading and Production of New Crust

- ❑ ***sea-floor spreading***
- ❑ ***mid-ocean ridges***

Subduction of Crust

- ❑ ***subduction zone***
- ❑ ***oceanic trenches***

The Formation and Breakup of Pangaea

Pre-Pangaea

Pangaea

Pangaea Breaks Up

Modern Continents Take Shape

The Continents Today

Plate Tectonics

- ❑ ***plate tectonics***

Plate Boundaries

Transform Faults

Earthquakes and Volcanoes

Hot Spots

- ❑ ***hot spots***

SUMMARY

Learning Activities and Critical Thinking

1. Select any five learning objectives from the list presented at the beginning of this chapter. Place the number selected in the space provided (no need to rewrite each objective). Using the following questions as guidelines only, briefly discuss your treatment of the objective.

- What did you know about the objective before you began?
- What was your plan to complete the objective?
- Which information source did you use in your learning (text, or other)?
- Were you able to complete the action stated in the objective? What did you learn?
- Are there any aspects of the objective about which you want to know more?

a)_____:____

____

____

____

____

____.

b)____ :____

____

____

____

____

c)____ :____

____

____

____

____

____

d)____ :____

____

____

____

____

____

e)____ :____

____

____

____

____

____

2. What are the three cycles that comprise the geologic cycle?

a)______________________________

b)______________________________

c)______________________________

3. Recreate on the following outline of Figure 8-1 (p. 253) the *geologic time scale.* Label each Eon, Era, Period, Epoch, and the date in millions of years before the present and note several of the important events in Earth's life history noted in the figure.

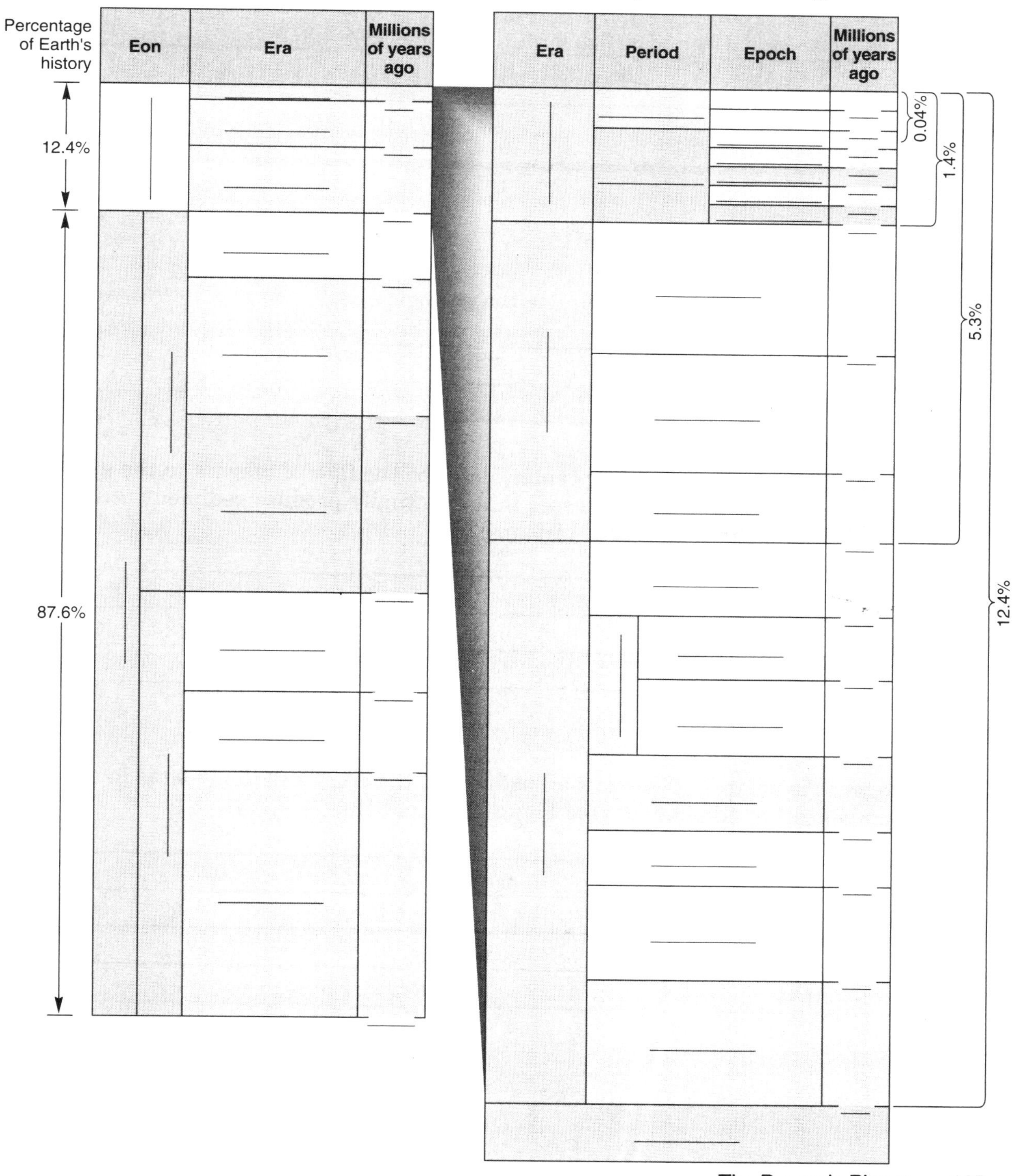

4. Using the sketch from Figure 8-3a, p. 255, detail the various layers within Earth. For each layer include its name and description (spaces on the left side) and depth below the surface in kilometers (spaces on the right side). The small wedge noted is detailed in the text in Figure 8-3b. You may want to use coloration with color pencils. Place your labels in the spaces provided. (Note: Earth's profile is placed on a map of the United States and Canada in Figure 8-4, p. 256; to give you a sense of scale.)

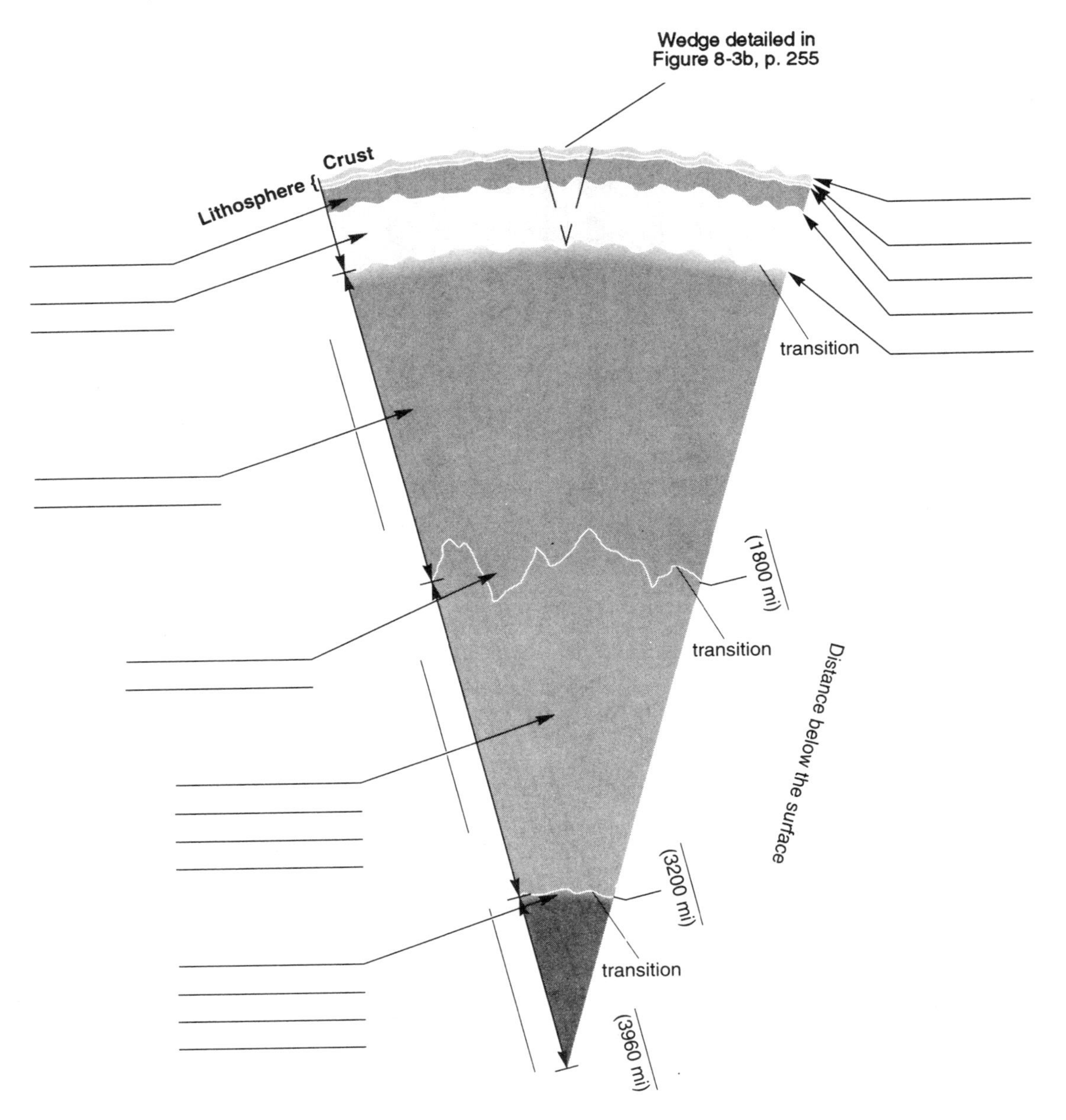

5. Beginning at a sea-floor spreading center, describe the flow of magma to the surface, its subsequent movements and the processes that eventually produce sediment accumulations (see the illustration in Figure 8-6, p. 260).

____

____

____.

6. Explain what is meant by: isostatic adjustment of the crust and the crust is in a constant state of compensating adjustment (see Figure 8-5, p. 258).

____

____

____.

7. Explain the generation of Earth's magnetic field. What relationship does it have to Earth's interior? What peculiar changes occur in the field that help scientists understand the crust?

____

____

____.

8. According to the text and illustrations (Figure 8-3, 255, Figure 8-5, p. 258, Figure 8-6, p. 260), use the space below to sketch a cross-section of continental and oceanic crust, Moho, uppermost mantle, and asthenosphere. Use coloration to highlight your sketch. Properly label each component.

9. Complete the labels on the following illustration identifying the various igneous forms and structures discussed in the text and Figure 8-9, p. 263.

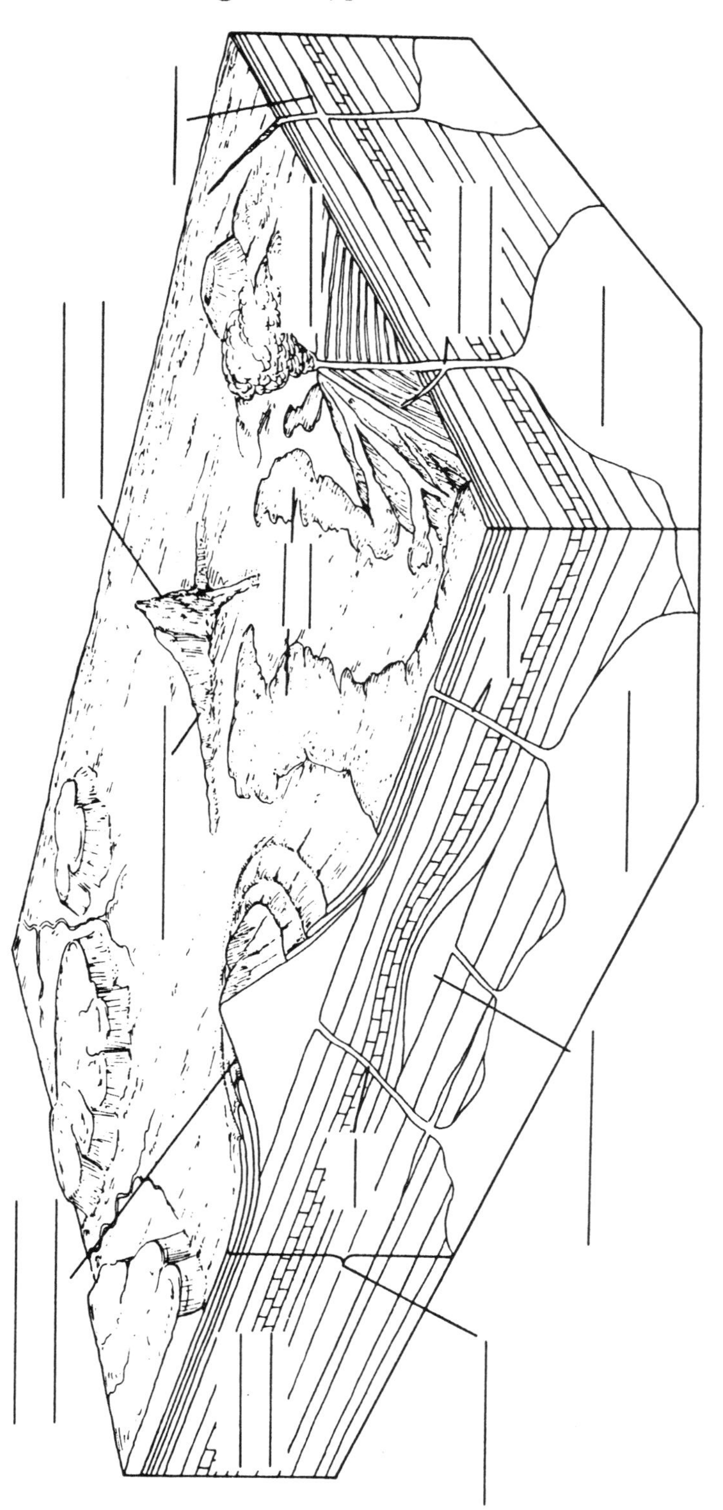

10. Overview the status of the various attempts to drill into the crust. Has anyone reached the Moho in drilling projects—the crust-mantle boundary? (See News Report #1, p. 259).

____

____

____

____.

11. List the eight natural elements most common in Earth's crust; include the percentage of each.

____

____

____.

12. Name the 14 major lithospheric plates illustrated in Figure 8-17 (p. 274). Add additional arrows along with the ones presented indicating the direction of plate movement. Place your labels and directional arrows on the map below. Lightly shading each plate with color pencils will make them easier to distinguish. Last: using red, shade those particular areas where earthquakes occur (see Figure 8-18, p. 275).

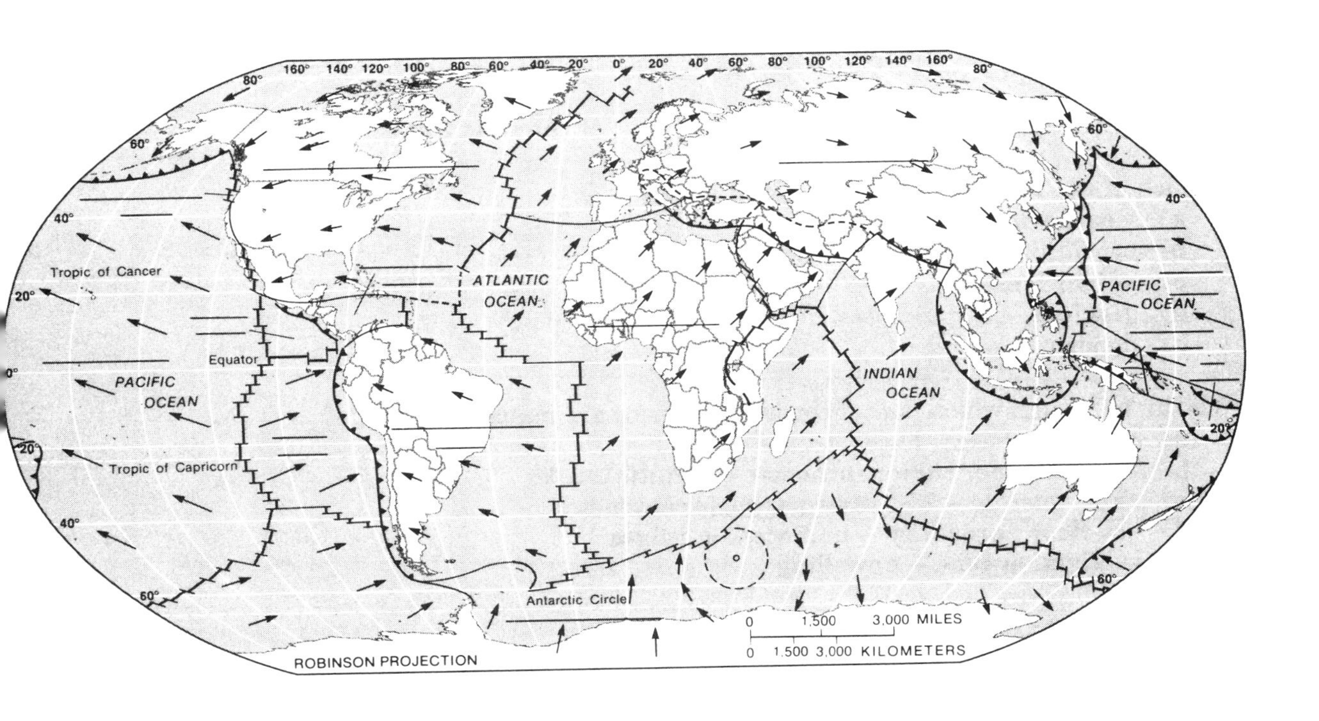

13. What does "Pangaea" mean? Who coined the name? Does it exist today? If, not explai briefly.

14. From the island of Hawaii to the island of Kauai, describe the ages of the islands fro youngest to oldest (Figure 8-19, p. 276). What processes explain this age distribution?

The age of Midway Island is:__________________.

15. How does News Report #3, "Yellowstone on the Move," relate to plate tectonics? Is there any similarity with the hot spot that produces the Hawaiian Islands? Explain and discuss.

16. Briefly describe coal (p. 265): originating material and formation processes, lithification and the four forms of coal, its use, associated problems, and the relationship between resource reserves and rates of consumption.

Sample Self-test

(Answers appear at the end of the study guide.)

1. Which of the following is endogenic in nature?
    a) weathering
    b) erosion
    c) volcanism
    d) glaciers
    e) deposition

2. Of the following, which pair of concepts or terms are matched correctly?
    a) asthenosphere – lower mantle
    b) extrusive igneous – granitic crust
    c) upper mantle – Earth's magnetic field
    d) sea-floor spreading – mid-oceanic ridges
    e) subduction zone – upwelling and sea-floor spreading

3. Earth's crust is roughly made up of
    a) mantle and core material
    b) at least 14 major plates capable of movement
    c) strong unbroken material
    d) a thick layer at least 300 kilometers deep
    e) a brittle material that does not move

4. According to your text, the layer within Earth that lies directly below the lithosphere, but above the upper mantle, is best described as
    a) resistant to movement of any type
    b) a granitic material that weighs an average of 2.7 grams per cm^3
    c) a liquid nickel-iron composition
    d) a plastic-like layer that shatters if struck but flows when subjected to heat and pressure–known as the asthenosphere

5. Continental drift is an earlier term that describes
    a) motions that occurred about two billion years ago
    b) crustal plate movements, proposed by a geographer in 1912
    c) a brittle crust incapable of movement
    d) an old theory that has been disproven

6. Which of the following supports the plate tectonics concept
    a) magnetic field patterns preserved in the rocks
    b) plant and animal fossil records
    c) radioactive decay dating of rocks on either side of a spreading centers
    d) all of the above are true
    e) none of these is valid, since the theory has fallen into disfavor and has been dropped

7. The deepest single group of features of Earth's crust (continental or oceanic) are
   a) the areas of plate subduction beneath the ocean
   b) the mid-ocean ridge systems
   c) the deep lakes in East Africa
   d) the abyssal plains

8. Igneous rocks are the product of
   a) the accumulation of pieces of preexisting rocks
   b) heat and pressure causing hardening and physical changes in the rock
   c) weathering and erosion processes
   d) solidifying and crystallizing magma

9. The basic premise of catastrophism is that "the present is the key to the past."
   a) true
   b) false

10. The cementation, compaction, and hardening of sediments is called lithification.
   a) true
   b) false

11. The oldest Earth rock is approximately 2.0 billion years old and was found in the Grand Canyon of Arizona.
   a) true
   b) false

12. The interior of Earth is known to science through direct physical observation and measurement in deep drill holes.
   a) true
   b) false

13. Earth's magnetic field is generated in the outer core and remains quite constant over time.
   a) true
   b) false

14. Seventy-five percent of Earth's crust is composed of only 2 elements–oxygen and silicon.
   a) true
   b) false

15. Rocks that solidify from a previous molten state are called metamorphic rocks.
   a) true
   b) false

16. Quartz (SiO_2) is generally higher in its resistance to weathering than are mafic minerals such as basalt.
   a) true
   b) false

17. Plate tectonics is regarded as the all-inclusive modern term for sea-floor spreading and subduction processes.

a) true
b) false

18. A convergent boundary characterizes the west coast of South America.

a) true
b) false

19. ________________________ assumes that *the same physical processes active in the environment today have been operating throughout geologic time*. The phrase "the ________________________ is the key to the ______________" is an expression coined to describe this principle. In contrast, the philosophy of ________________________ that attempts to fit the vastness of Earth's age and the complexity of its rocks into a shortened time span.

20. A ________________ is an element or combination of elements that forms an inorganic natural compound. Of the nearly 3000 minerals, only 20 are common, with just 8 of those making up _______% of the minerals in the crust. A ______________ is an assemblage of minerals bound together or an aggregate of pieces of a single mineral.

Name:________________________________ Class Section:________________

Date:__________________ Score/Grade:______________

Earthquakes and Volcanoes

Chapter Overview

Tectonic activity has repeatedly deformed, recycled, and reshaped Earth's crust during its 4.6 billion year existence. The principal tectonic and volcanic zones lie along plate boundaries or in areas where Earth's crustal plates are affected by processes in the asthenosphere. The arrangement of continents and oceans, the origin of mountain ranges, and the locations of earthquake and volcanic activity are all the result of these dynamic endogenic processes.

The chapter opening illustration, "The Ocean Floor," is a bridge between Chapter 8 and this chapter. The plate map in Figure 8-17 (p. 274) and the map of earthquake and volcanic occurrence in Figure 8-18 (p. 275), as well the world structural map in Figure 9-13 (p. 296), all correlate within this opening illustration. Follow the tour of the ocean floor on page 282 in the text and see if you can identify the features and places described.

Learning Objectives

The following learning objectives help guide your reading and comprehension efforts. The operative word is in *italics*. Read and work with these carefully and note that exercise #1 asks you about five of these objectives. After reading the chapter and using this workbook, you should be able to:

1. *Analyze* the chapter opening map of "The Ocean Floor," and *identify* the elements of plate tectonics in the illustration.
2. *Portray* first, second, and third orders of relief and *relate* examples of each.
3. *Name* and *define* Earth's major topographic regions.
4. *List* the three general regions of mountain building and uplifted crust.
5. *Describe* the variable makeup of continental crust.
6. *Define* displaced terranes and *relate* them to the accretion and growth of continental platforms.
7. *Describe* and *contrast* crustal deformation: compression, tension, and shearing.
8. *Diagram* folding processes and *label* anticlinal and synclinal features.
9. *Describe* the faulting process and *relate* it to the occurrence of earthquakes.
10. *List* four principal types of faults and *identify* related motions and landform features.
11. *Describe* a horst and graben landscape; *describe* a tilted-fault block.
12. *Define* an orogeny (orogenesis) and *identify* the two main orogenic belts (mountain chains) on Earth.

13. *Relate* the three types of plate collisions associated with orogenesis and *use* specific examples of each.
14. *Explain* the circum-Pacific belt (ring of fire) in relation to plate tectonics.
15. *Review* the basic world structural regions.
16. *Contrast* the two earthquake events that occurred in China in 1975 and 1976.
17. *Distinguish* between the Richter scale and the Mercalli scale and *relate* the number of expected occurrences per year at varying levels of magnitude.
18. *Define* the elastic-rebound theory and *relate* the role of asperities and asperity breaks.
19. *Relate* the 1992 and 1994 earthquakes in Southern California to the San Andreas fault system and the principles or other fault systems beneath Los Angeles discussed in this chapter.
20. *Describe* paleoseismology and *relate* the possible role it may play in earthquake forecasting.
21. *Distinguish* between an effusive and an explosive eruption and *describe* related landforms.
22. *Explain* the relationship of the source of the magma and the type of volcanic eruption that occurs.
23. *Review* the Mount Saint Helens eruption sequence and its consequences.

Outline Headings and Glossary Review

These are the first- second-, and third-order headings that divide this chapter. The key terms and concepts that appear **boldface** in the text are listed under their appropriate heading in bold italics; these highlighted terms appear in the text glossary. A check-off box is placed next to each key term so you can mark your progress through the chapter as you define these in your reading notes or prepare note cards.

The Ocean Floor

Earth's Surface Relief Features

❑ ***relief***
❑ ***topography***

Crustal Orders of Relief

First Order of Relief

❑ ***continental platforms***
❑ ***ocean basins***

Second Order of Relief

Third Order of Relief

❑ ***hypsometry***

Earth's Topographic Regions

Crustal Formation Processes

Continental Shields

❑ ***continental shield***

Building Continental Crust

Terranes

❑ ***terranes***

Crustal Deformation Processes

Folding

❑ ***folding***
❑ ***anticline***
❑ ***syncline***

Faulting

❑ ***faulting***
❑ ***earthquake***
❑ ***normal fault***
❑ ***reverse fault***
❑ ***thrust fault***
❑ ***strike-slip fault***
❑ ***horst***
❑ ***graben***

Orogenesis (Mountain Building)

❑ ***orogenesis***
❑ ***tilted fault-block***

Types of Orogenies

❑ ***circum-Pacific belt***

The Appalachian Mountains

World Structural Regions

❑ ***Cordilleran system***

Earthquakes

Earthquake Essentials

❑ ***seismograph***
❑ ***Richter scale***
❑ ***epicenter***

The Nature of Faulting

❑ ***elastic-rebound theory***

The San Francisco Earthquakes

The Southern California Earthquakes

Earthquake Forecasting, Preparedness, and Planning

Volcanism

❑ ***volcano***
❑ ***crater***
❑ ***geothermal energy***
❑ ***lava***
❑ ***tephra***
❑ ***cinder cone***
❑ ***caldera***

Locations of Volcanic Activity

Types of Volcanic Activity

Effusive Eruptions

❑ ***effusive eruption***
❑ ***shield volcano***
❑ ***plateau basalts***

Explosive Eruptions

❑ ***explosive eruption***
❑ ***composite volcano***

SUMMARY

Learning Activities and Critical Thinking

1. Select any five learning objectives from the list presented at the beginning of this chapter. Place the number selected in the space provided (no need to rewrite each objective). Using the following questions as guidelines only, briefly discuss your treatment of the objective.

- What did you know about the objective before you began?
- What was your plan to complete the objective?
- Which information source did you use in your learning (text, or other)?
- Were you able to complete the action stated in the objective? What did you learn?
- Are there any aspects of the objective about which you want to know more?

a)_____:____

____

____

____

____

____.

b)_____:____

____

____

____

____

____.

c)____ :____

____

____

____

____

____.

d)____ :____

____

____

____

____

____.

e)____ :____

____

____

____

____

____.

2. Define each order of relief and give a specific example of each.

a) first order:____

____.

b) second order:____

____.

c) third order:____

____.

3. In general terms describe the features of each type of topographic region mapped in Figure 9-2, p. 284.

a) plain:____

____.

b) high tableland:____

____.

c) hills and low tablelands:____

____.

d) mountains:____

____.

e) widely spaced mountains:____

____.

4. List Earth's nine major continental shields labeled in Figure 9-3, p. 285:______________

____

____.

5. In the following illustration from Figure 9-5a, p. 288, identify each of the components of this folded landscape. Circle the portion of the illustration that you think is depicted in the photograph in Figure 9-5b.

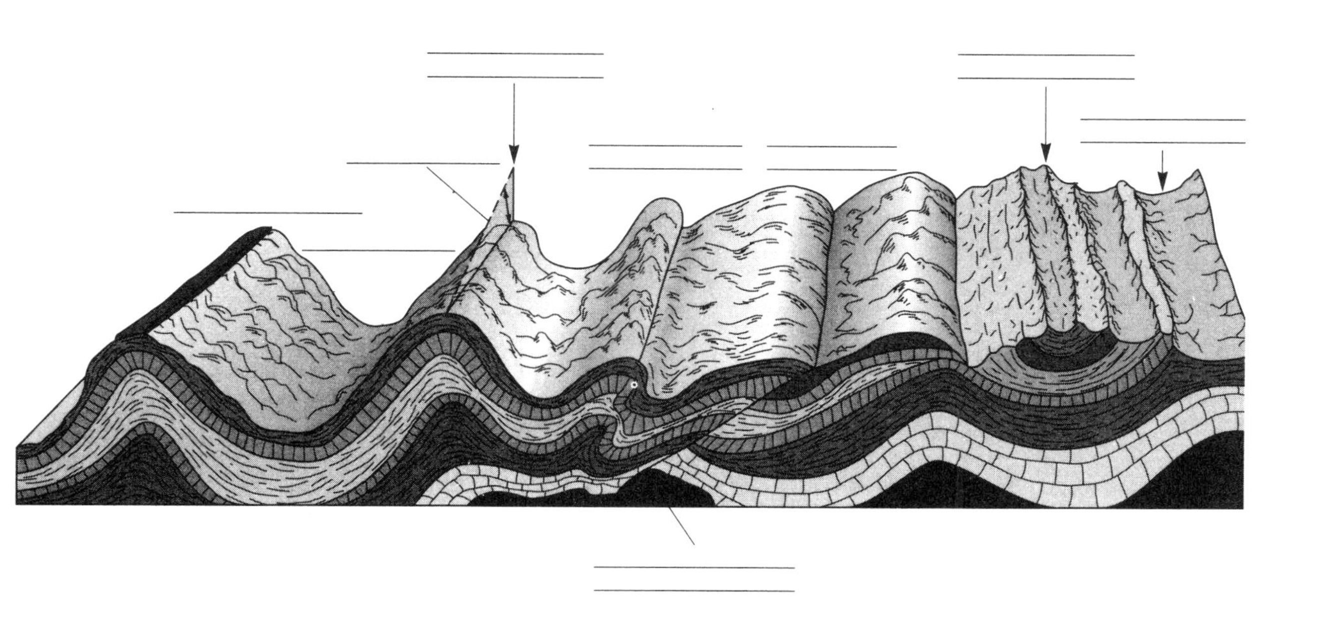

6. For Figure 9-8, p. 291, identify each of the components of these fault types and add proper labels.

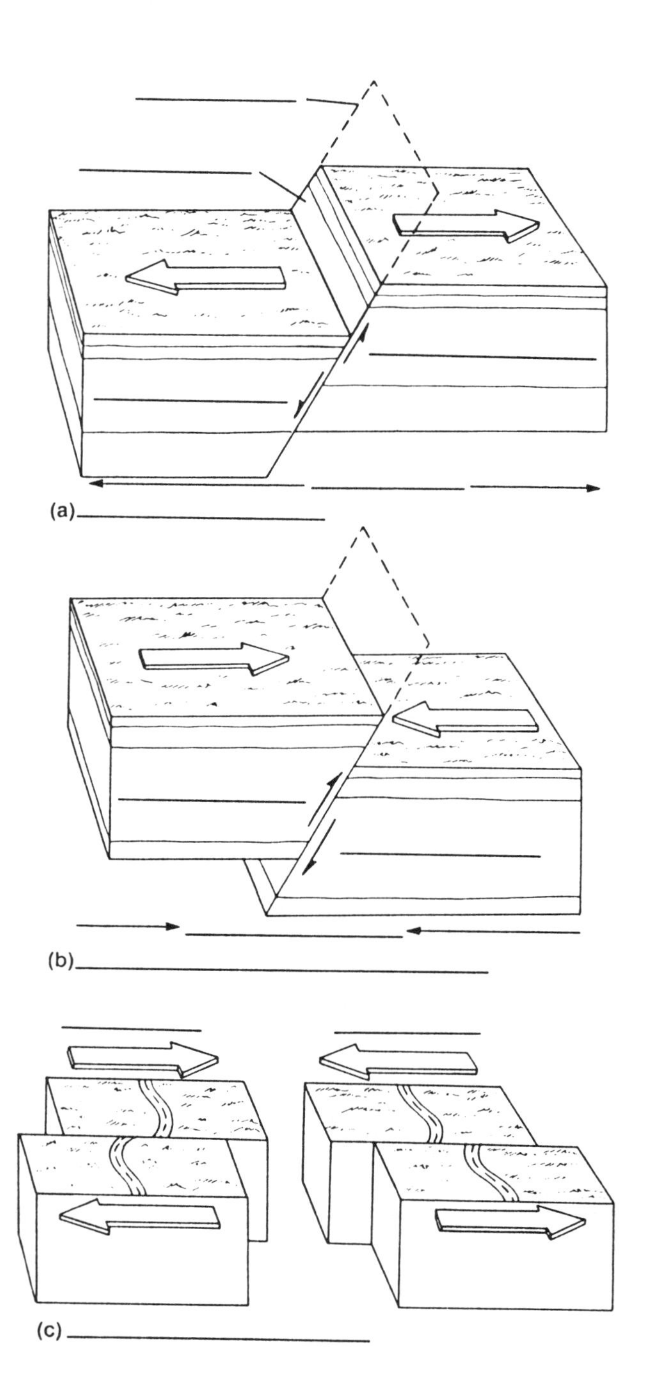

7. What are the three types of plate convergence mentioned in the text? What specific features do you associate with each? Describe the examples given in Figure 294, p. 294, that each leader line points to the ocean floor inset maps.

(a)____

____.

(b)____

____.

(c)____

____.

8. Describe how earthquakes are measured and characterized in terms of intensity and magnitude (page 297-98.

(a) intensity:____

____

____.

(b) magnitude:____

____

____.

9. In terms of earthquake characteristics and frequencies expected each year complete the following from Table 9-1, p. 297.

Characteristic Effects in Populated Areas	Approximate Intensity (modified Mercalli scale)	Approximate Magnitude (Richter scale)	Number per year
Nearly total damage			
Great damage			
Considerable-to-serious			
Felt-by-all			
Felt-by-some			
Not felt, but recorded			

10. Assuming that the aerial photograph below was taken looking south, label the various physical aspects of plate tectonics and faulting shown in the photo. Labels include: ❑ the Pacific plate, ❑ the North American plate, ❑ the rift along the San Andreas fault, and in the two lower boxes place ❑ the directional arrows showing the relative motion of the plates. (Disregard the fence line on the edge of the valley that runs along the east side of the hills.)

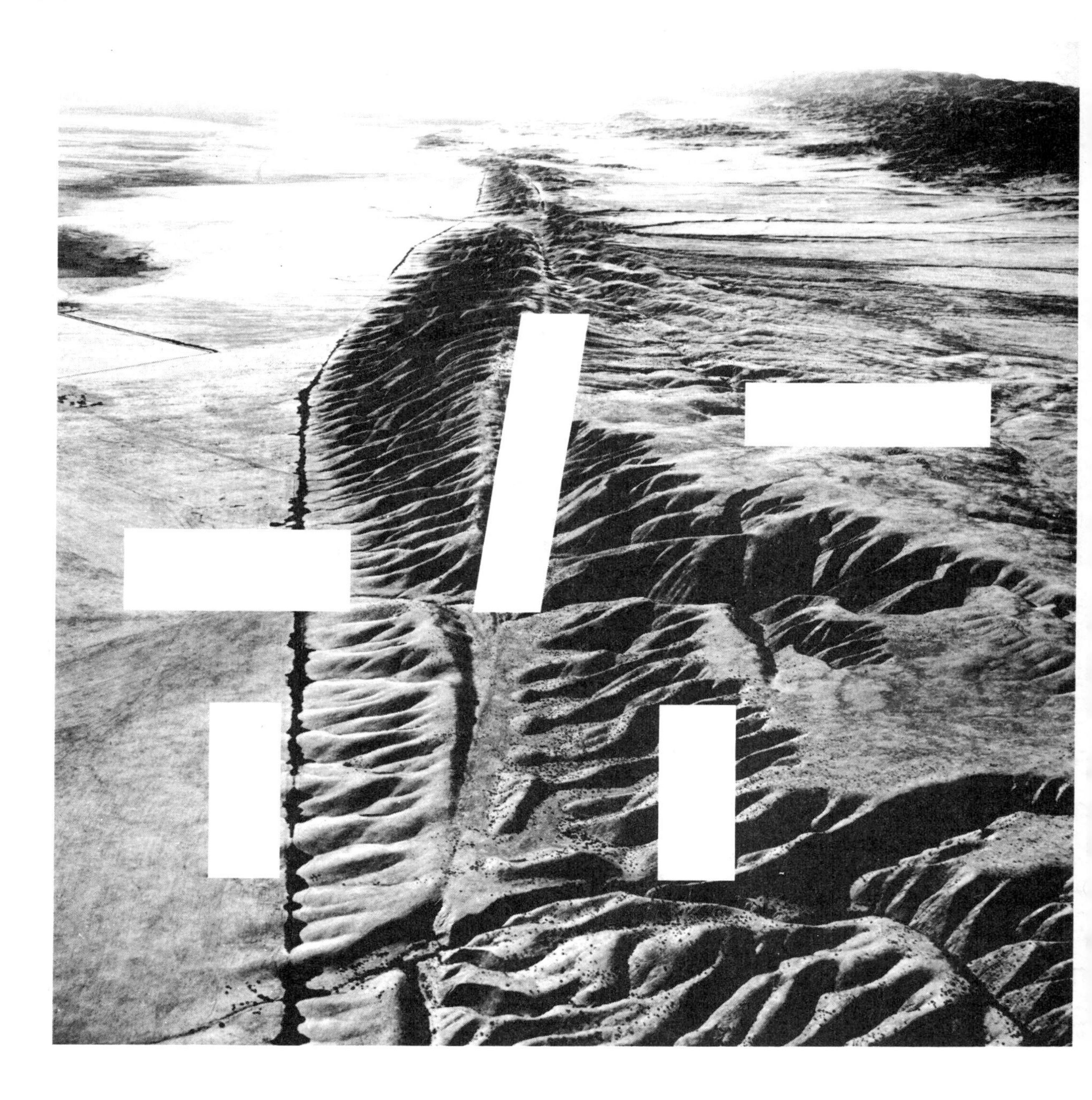

11. From the description in the text and Figure 9-15, p. 301, describe how the thrust action of the Loma Prieta earthquake was generally <u>not characteristic</u> of the San Andreas fault system, even though it was produced along a portion of it.

____

____

____.

12. Relative to the San Andreas fault system define each of the following processes or motions (pp. 298-99).

a) transform faults:____

____.

b) strike-slip:____

____.

c) right-lateral:____

____.

13. The text uses a simplified criteria to classify volcanic eruptions—*effusive eruptions* and *explosive eruptions*. Describe the distinction between these two forms of volcanism, including:

❑ the origin of magma, ❑ eruption characteristics, and ❑ landform features produced.

a) effusive eruptions:____

____

____

____

____

____.

b) explosive eruptions:____

____

____

____

____

____.

14. The principal mechanisms of volcanic activity are illustrated in Figure 9-17 (p. 304-05), which is reproduced here (in part). Identify aspects of volcanic activity by placing the proper terms as labels in the spaces provided. Use coloration to highlight key aspects.

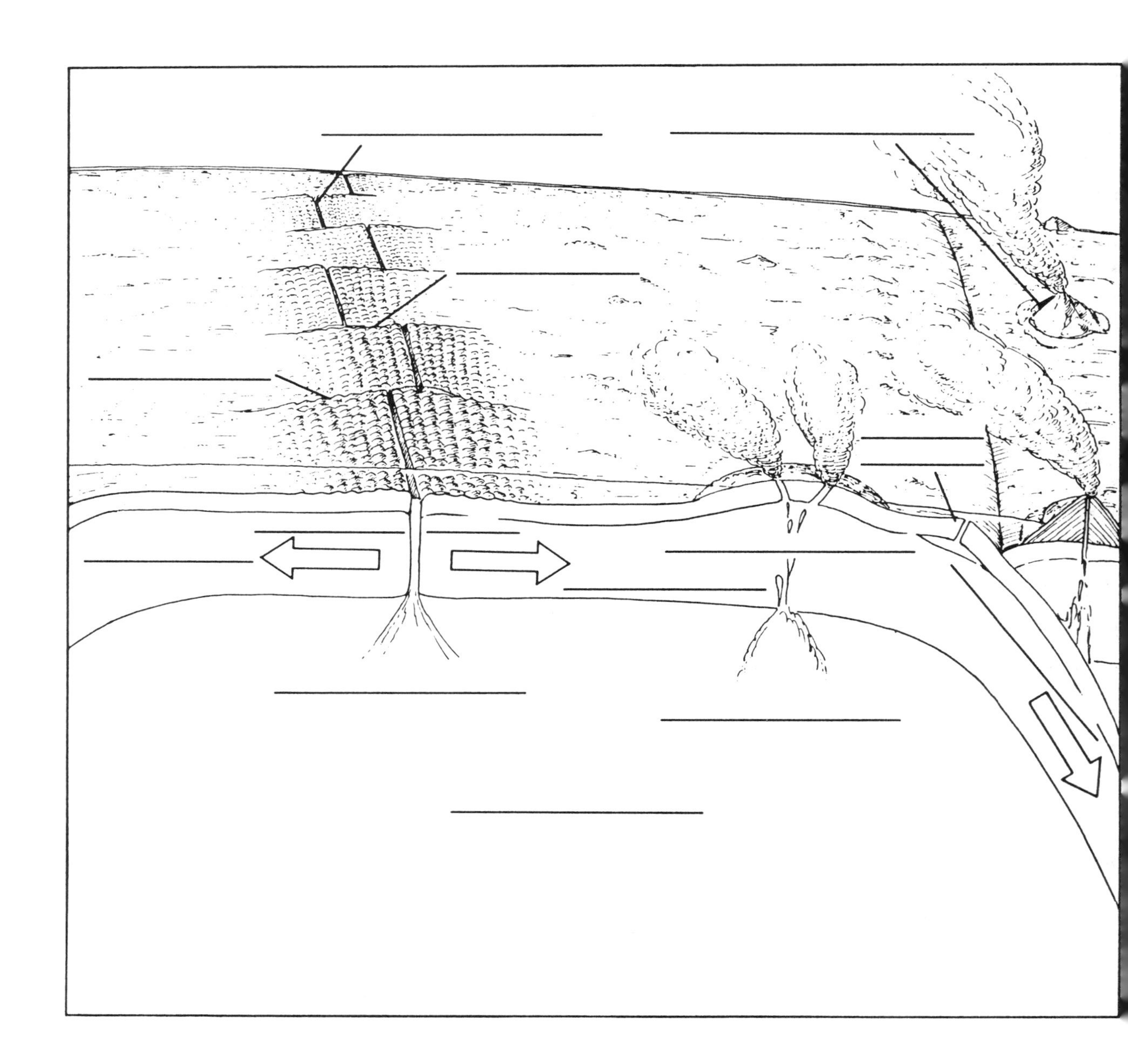

15. Given the discussion of plate tectonics, the crust and Moho, and earthquakes in these two chapters, briefly discuss and explain "Damage Strikes at Distance from Epicenters" (News Report #2, p. 299.

____

____

____

____

____.

Sample Self-test

(Answers appear at the end of the study guide)

1. The deepest single group of features on the ocean floor are
   a) oceanic trenches
   b) mid-ocean ridge systems
   c) continental areas
   d) areas beneath the ice sheets

2. Which of the following represents a second order of relief?
   a) the North American plate
   b) the hills and valleys
   c) the Alps and the Rockies
   d) the oceanic basins

3. Terranes refer to
   a) the topography of a tract of land
   b) the "lay of the land"
   c) displaced and migrating pieces of Earth's crust
   d) features in the formation of Europe but not North America
   e) the continental shields within each continent

4. Faults that result from vertical displacements caused by crustal rifting of one portion of Earth's crust are called
   a) strike-slip, or transform faults
   b) reverse faults
   c) folded faults
   d) normal faults and reverse faults

5. Hawaiian volcanoes are
   a) explosive in nature
   b) related to subduction zones
   c) effusive and shield-like
   d) presently dormant

6. The damage caused by the Loma Prieta earthquake
   a) was associated with a normal fault and not the San Andreas fault
   b) involved a portion of the San Andreas Fault in a strike-slip motion at depth
   c) was related to severe surface faulting and breaking
   d) was isolated to near the epicenter northeast of Santa Cruz

7. Mount Saint Helens is a type of composite volcano associated with a subducted oceanic plate.
   a) true
   b) false

8. The most active tectonic regions of North and South America are on the
   a) western coasts
   b) eastern coasts
   c) southern coasts
   d) mid-continent

9. An instrument used to record vibrations in the crust is the
   a) land barometer
   b) seismograph
   c) the Richter scale
   d) tiltmeter

10. According to the text, the major impact of an earthquake prediction on a region would be
    a) a decline of and damage to the local economy
    b) increased business activity
    c) a slight but general increase in property values
    d) upward changes in employment opportunities

11. Which of the following match-ups is incorrect?
    a) anticline – a trough
    b) normal fault – horizontal motion
    c) reverse fault – overlying blocks that move upward relative to the footwall block
    d) syncline – a ridge

12. The ocean floor still remains a complete mystery to modern science.
    a) true
    b) false

13. The Hawaiian Islands mark one end of a hot-spot track that stretches to Alaska.
    a) true
    b) false

14. Plains are identified as surfaces with local relief that does not exceed 100 m (325 ft).
    a) true
    b) false

15. Continental crust contains higher amounts of silica and aluminum and is composed of magma that is quite viscous when compared with magma directly from the asthenosphere.
   a) true
   b) false

16. Folding is principally associated with earthquake activity.
   a) true
   b) false

17. The Sierra Nevada and Grand Tetons are examples of tilted fault-block landscapes.
   a) true
   b) false

18. The Appalachian Mountain region is an example of a folded landscape.
   a) true
   b) false

19. A continental–continental collision produced over 1000 km of crustal shortening where India and the Asian landmass collided.
   a) true
   b) false

20. Earthquake damage may occur at a distance from the epicenter, as was the case in the Mexico earthquake in 1985 and the Loma Prieta quake in 1989.
   a) true
   b) false

Name:________________________________ Class Section:__________________

Date:____________________ Score/Grade:______________

Weathering, Karst Landscapes, and Mass Movement

10

Chapter Overview

Chapter 10 begins the treatment of the external processes that affect Earth's surface. As the landscape is formed, a variety of processes simultaneously operate to wear it down, all under the influence of gravity. The dynamics of slope and ever-changing adaptations and conditions of various slope elements produce a dynamic equilibrium among rock structure, climate, local relief, and elevation. Physical and chemical weathering operate toward the overall reduction of the landscape and the release of essential minerals from bedrock. In addition, mass movement of surface material rearranges landforms, providing often-dramatic reminders of the power of nature. The complete outline headings for this chapter are:

Students can be asked if they have noticed highways in mountainous and cold climates that appear rough and broken. Roads that experience freezing weather seem to pop up in chunks each winter. Or, maybe they have seen older marble structures, such as tombstones, etched and dissolved by rainwater. Similar physical and chemical weathering processes are important to the overall reduction of the landscape and the release of essential minerals from bedrock. A simple examination of soil gives evidence of weathered mineral grains from many diverse sources. In addition, mass movement of surface material rearranges landforms, providing often-dramatic reminders of the power of nature.

Learning Objectives

The following learning objectives help guide your reading and comprehension efforts. The operative word is in *italics*. Read and work with these carefully and note that exercise #1 asks you about five of these objectives. After reading the chapter and using this workbook, you should be able to:

1. *Define* the science of geomorphology.
2. *Describe* various denudation processes and *relate* specific examples of each.
3. *Define* the base level concept and *distinguish* between it and local base levels.
4. *Contrast* the evolutionary geomorphic cycle model with the dynamic equilibrium approach to slopes and landforms.

5. *Illustrate* the forces at work on materials residing on a slope.
6. *Define* weathering and *distinguish* physical from chemical weathering processes.
7. *Portray* the relationship between bedrock, regolith, parent materials, and any soils that might exist.
8. *Explain* what is meant by differential weathering and erosion and *illustrate* it with examples.
9. *Describe* crystallization as a physical weathering process.
10. *Relate* the process of hydration to rock disintegration.
11. *Explain* the role of freezing water as a physical weathering agent.
12. *Analyze* the process that forms domelike features on a granitic batholith.
13. *Describe* the susceptibility of different minerals to the chemical weathering process called hydrolysis.
14. *Outline* the various processes and features associated with karst topography.
15. *Describe* the relationship between the Arecibo radio telescope in Puerto Rico and karst landscape features.
16. *Differentiate* between mass movement and mass wasting and *relate* examples of each.
17. *Analyze* the events that took place along the Madison River Canyon near West Yellowstone in August 1959.
18. *Describe* the events that took place on the west face of Nevado Huascarán, Peru, in 1962 and 1970.
19. *Portray* the various types of mass movements and *identify* examples for each in relation to moisture content and speed of movement.
20. *Explain* the process of soil creep.
21. *Explain* the impact of human-induced mass movements.

Outline Headings and Glossary Review

These are the first- second-, and third-order headings that divide this chapter. The key terms and concepts that appear **boldface** in the text are listed under their appropriate heading in bold italics; these highlighted terms appear in the text glossary. A check-off box is placed next to each key term so you can mark your progress through the chapter as you define these in your reading notes or prepare note cards.

Landmass Denudation

❑ ***geomorphology***
❑ ***denudation***

Base Level of Streams

base level

Dynamic Equilibrium Approach to Landforms

❑ ***dynamic equilibrium model***
❑ ***geomorphic threshold***

Slopes

❑ ***slopes***

Weathering Processes

❑ ***weathering***
❑ ***regolith***
❑ ***bedrock***
❑ ***sediment***
❑ ***parent material***
❑ ***joints***
❑ ***differential weathering***

Physical Weathering Processes

❑ ***physical weathering***

Crystallization

Hydration

❑ ***hydration***

Frost Action

❑ ***frost action***

Pressure-Release Jointing

❑ ***sheeting***
❑ ***exfoliation dome***

Chemical Weathering Processes

❑ ***chemical weathering***
❑ ***spheroidal weathering***

Hydrolysis

❑ ***hydrolysis***

Oxidation

❑ ***oxidation***

Carbonation and Solution

❑ ***solution***
❑ ***carbonation***

Karst Topography and Landscapes

❑ ***karst topography***
❑ ***sinkholes***

Caves and Caverns

Mass Movement Processes

Mass-Movement Mechanics

❑ ***mass movement***

The Role of Slopes

Classes of Mass Movements

Falls and Avalanches

❑ ***rockfall***
❑ ***debris avalanche***

Slide

❑ ***landslide***

Flow

Creep

❑ ***soil creep***

Human-Induced Mass Movements

❑ ***scarification***

SUMMARY

Learning Activities and Critical Thinking

1. Select any five learning objectives from the list presented at the beginning of this chapter. Place the number selected in the space provided (no need to rewrite each objective). Using the following questions as guidelines only, briefly discuss your treatment of the objective.

- What did you know about the objective before you began?
- What was your plan to complete the objective?
- Which information source did you use in your learning (text, or other)?
- Were you able to complete the action stated in the objective? What did you learn?
- Are there any aspects of the objective about which you want to know more?

a)____:____

____

____

____

____

____.

b)____:____

____

____

____

____

____.

c)____ :____

____

____

____

____

____.

d)____ :____

____

____

____

____

____.

e)____ :____

____

____

____

____

____.

2. Differentiate between the *ultimate base level* and a *local base level*. Use examples to illustrate your descriptions.

a) ultimate base level:____

____

____

____.

b) local base level:____

____

____

____.

3. Using the following outline of Figure 10-3a and b, p. 319, complete the labels and components noted with leader lines detailing the forces acting on a slope and the principal elements of slope form.

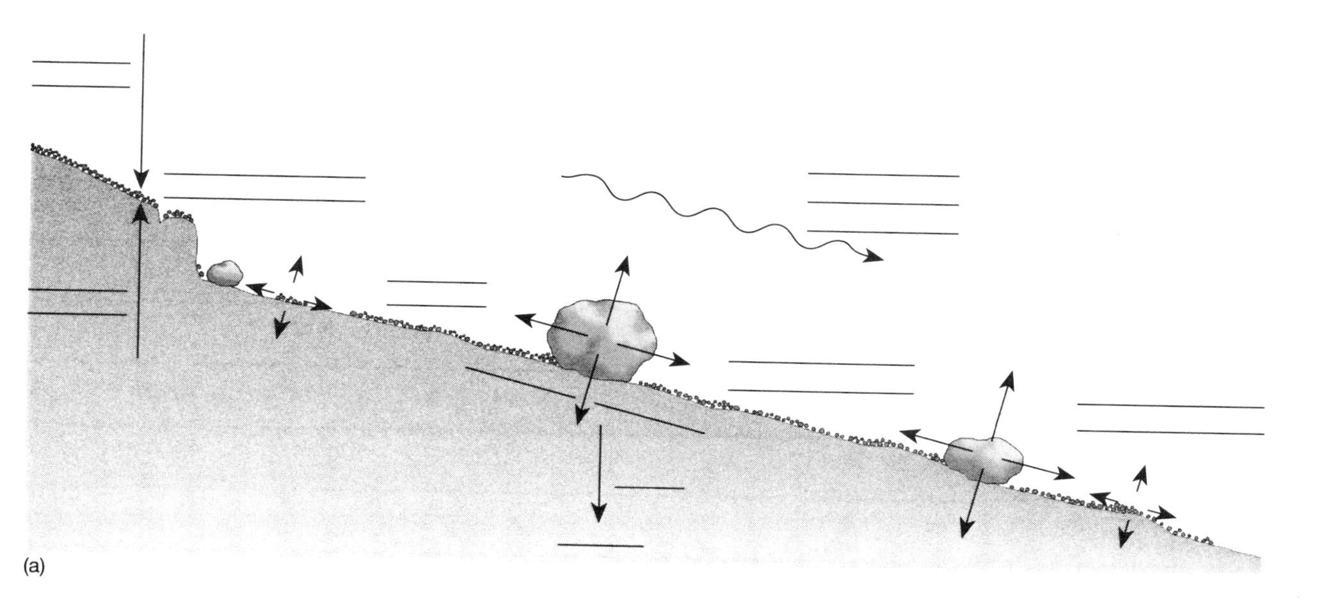
(a)

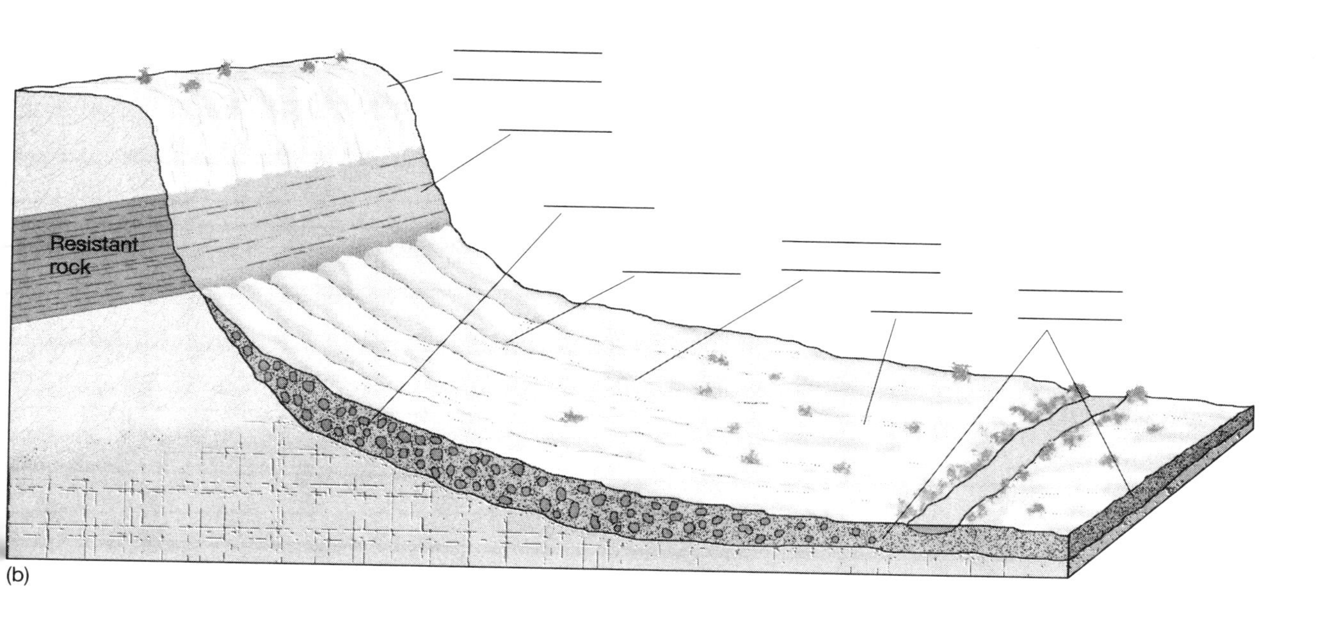

(b)

4. Differentiate between *bedrock*, *regolith*, and *soil*. You may want to refer ahead to pag
454 when composing your soil response.

(a) bedrock:

(b) regolith:

(c) soil:

5. Explain *pressure-release jointing* and the processes of *sheeting* and *exfoliation* as types of *physical weathering*. Can you identify these processes and features in the two photographs in Figure 10-9, p. 324?

6. Differentiate between *physical weathering* and *chemical weathering* processes. Give examples of each in the spaces provided.

(a) physical weathering:

example:

(b) chemical weathering:

example:

7. Review the photograph in Figure 10-10, p. 325, and describe the features and processes that you identify operating in and on the rock formation. Review the caption and the text section in preparing your response.

____

____

____.

8. Answer the following relative to Figure 10-11 a and b, p. 327.

(a) What factors explain the pitted texture and indentations in the landscape?__________

____.

(b) What do you think happened to the disappearing streams noted in the illustration?

____

____.

(c) Describe the processes at work that produced the sinkhole pictured in Florida.________

____

____.

9. How is the construction of the Arecibo radio telescope observatory (Figure 10-12, p. 328) related to the characteristics of karst features? Explain.____________________

____

____.

10. Relative to caves and caverns, define the following features and related processes that produce them:

(a) dripstone:______________________________________

____.

(b) stalactites:_____________________________________

____.

(c) stalagmites:____________________________________

____.

(d) column:____

____.

(e) Can you identify any of these four features in Figure 10-14, p. 329? ______________________

____

11. Identify with labels and descriptions the features of an underground cave as illustrated in Figure 10-15, p. 330.

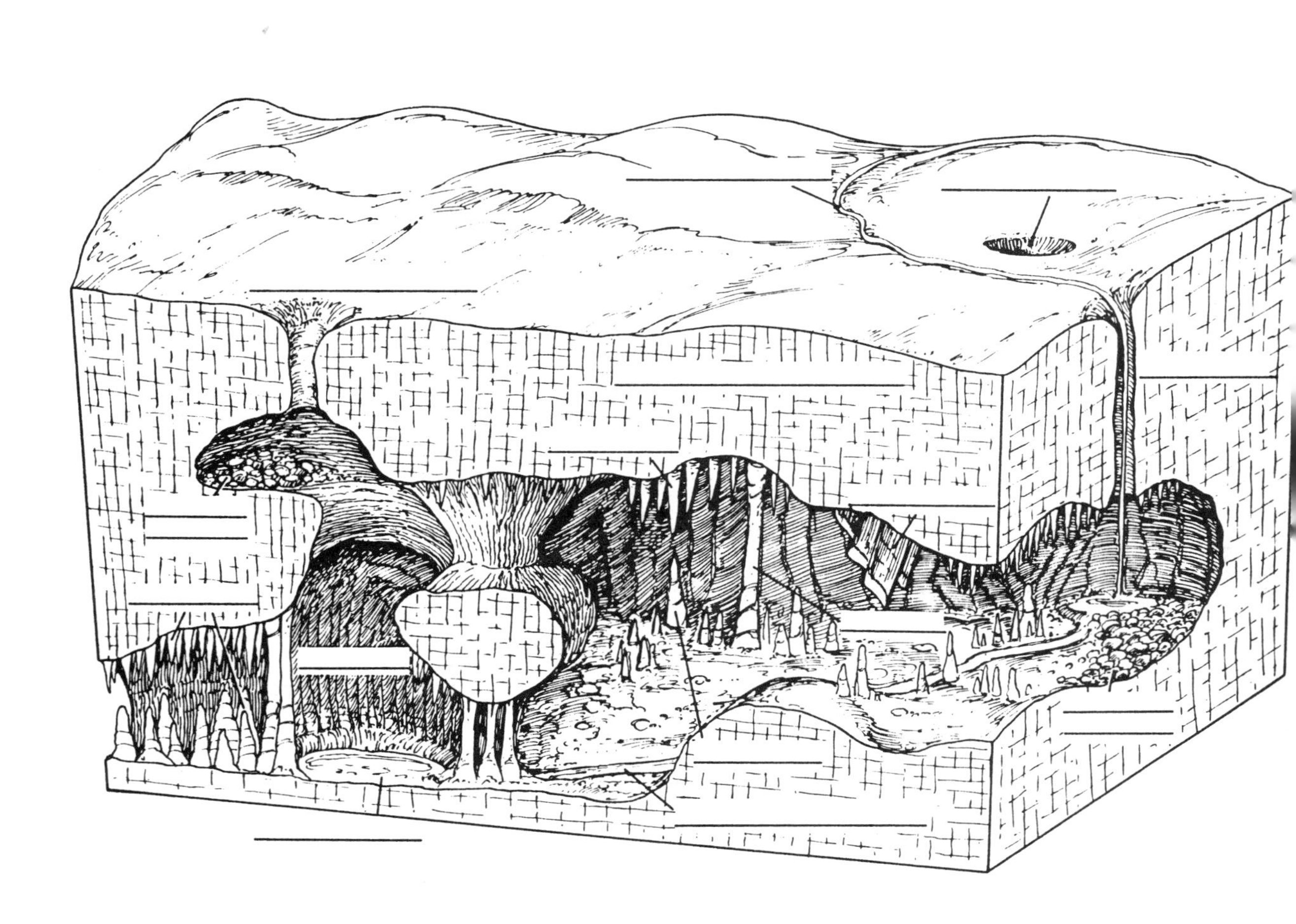

12. List the physical factors that contributed to the landslide event that occurred along the Madison River in 1959 (Figure 10-16, pp. 332-33). Include the triggering mechanism in your discussion.____

____

____

____.

13. Mass movement and mass wasting events are differentiated on the basis of water content and rates of movement. The following illustration is derived from Figure 10-17, p. 333. Identify the type of movement depicted in each example and label them in the spaces provided.

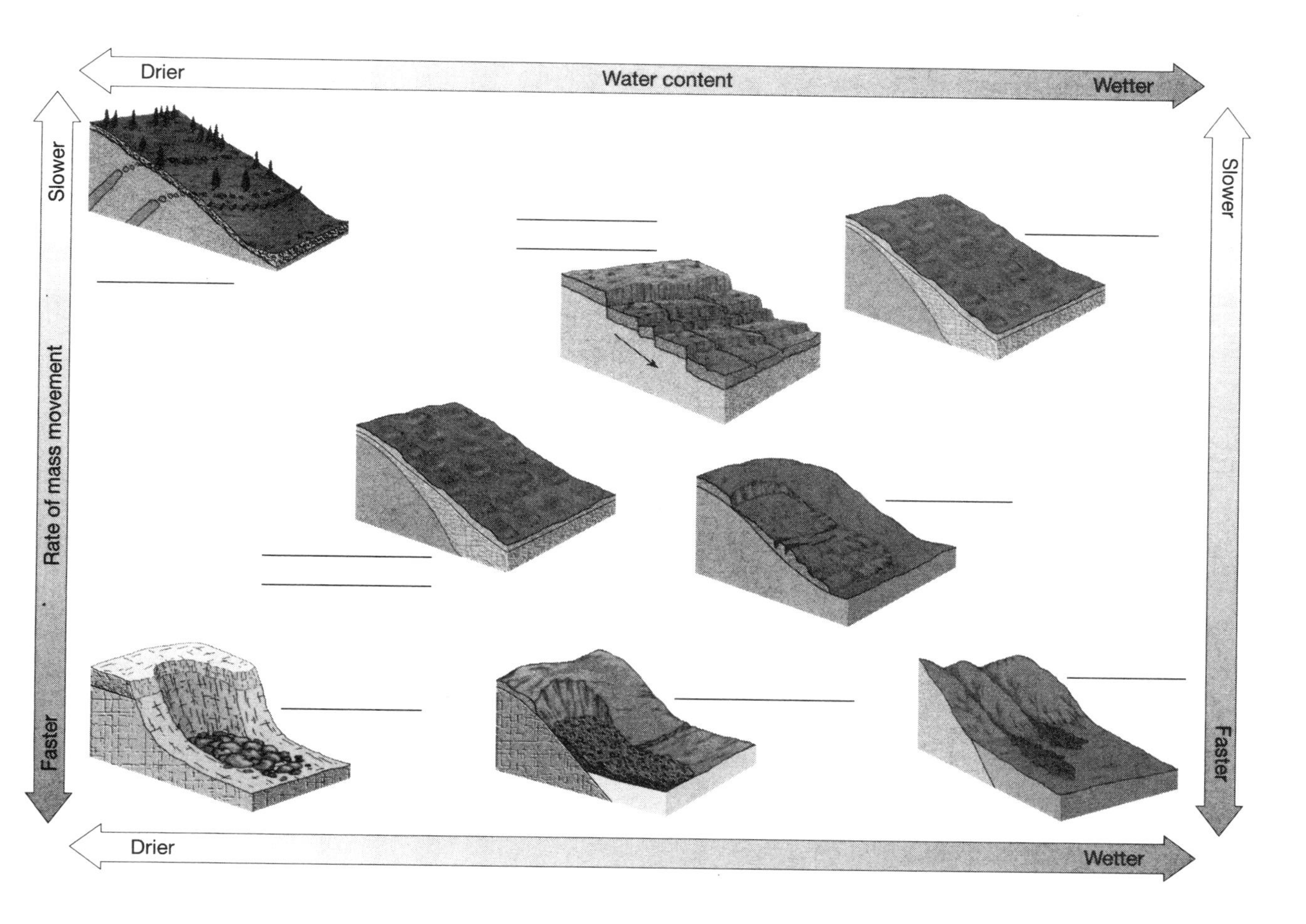

Sample Self-test

(Answers appear at the end of the study guide.)

1. The science that specifically studies the origin, evolution, form, and spatial distribution of landforms is
   a) geology
   b) geography
   c) geomorphology
   d) environmental chemistry
2. All processes that cause reduction and rearrangement of landforms are included in the term
   a) mass movement
   b) mass wasting
   c) weathering
   d) denudation

3. John Wesley Powell put forward the idea of base level which refers to
   a) an evolutionary cycle of landscape development
   b) an imagined surface that extends inland from sea level, inclined gently upward
   c) a level below which a stream cannot erode its valley
   d) flat plateaus
   e) both b) and c) are correct

4. The dynamic equilibrium model refers to
   a) a balancing act between tectonic uplift and reduction rates of weathering and erosion in a given landscape
   b) a theory involving the cyclic or evolutionary development of a landscape
   c) landscapes that do <u>not</u> evidence on-going adaptations to ever-changing conditions
   d) an important concept first stated by William Morris Davis

5. The disintegration and dissolving of surface and subsurface rock is called
   a) erosion
   b) mass wasting
   c) landmass denudation
   d) weathering

6. Chemical weathering is greatest under conditions of
   a) higher mean annual rainfall and temperatures
   b) lower mean annual rainfall and higher mean annual temperature
   c) lower mean annual rainfall and lower temperatures
   d) temperatures below freezing

7. An exfoliation dome in granitic rock forms through a process known as
   a) pressure-release jointing
   b) hydrolysis
   c) crystallization
   d) freeze-thaw action

8. Hydration is a process whereby a mineral
   a) chemically combines with water in chemical reactions
   b) dissolves in the presence of a weak acid
   c) swells upon the absorption of water, creating stress in rock
   d) oxidizes, most familiar in the "rusting" of iron

9. A local base level could be formed by a reservoir and the presence of a dam.
   a) true
   b) false

10. The geomorphic threshold is reached when a landform system possesses enough energy to overcome resistance against movement.
   a) true
   b) false

11. Slopes principally act as open material systems.
   a) true
   b) false

12. Regolith refers to partially weathered bedrock and may be missing or undeveloped in a region.
   a) true
   b) false

13. Freeze-thaw action of water in rocks is related to hydrogen bonding between water molecules and the expansion of water by as much as 9% in its volume as it cools and freezes.
   a) true
   b) false

14. Exfoliation domes and sheeting represent a form of pressure-release jointing.
   a) true
   b) false

15. Debris avalanches devastated the same villages in Peru twice within an eight-year period.
   a) true
   b) false

16. A lahar is a form of mass movement associated with snow avalanches.
   a) true
   b) false

17. The science that describes the origin, evolution, form, and spatial distribution of landforms is

______________________________________.

18. Limestone is so abundant on Earth that many landscapes are composed of it. These landscape comprise_____________________topography originally named for the ____________Plateau in Yugoslavia, where these processes were first studied.

19. A persistant mass movement of surface soil is called ____

20. Landscapes that result from human-induced mass movements are sometimes described by th term__________________________________.

Name:______________________________ Class Section:________________

Date:________________ Score/Grade:____________

Rivers and Related Landforms

11

Chapter Overview

This chapter begins with a discussion of the drainage basin which is a basic hydrologic unit. With this established, streamflow characteristics, gradient, and deposition are presented as water cascades through the hydrologic system. The human component is irrevocably linked to streamflow as so many settlements are along river banks and on floodplains. Not only do rivers provide us with essential water supplies, but they also receive, dilute, and transport wastes and provide critical cooling water for industry. Rivers have been of fundamental importance throughout human history. This chapter discusses the dynamics of river systems and their landforms.

A daily check of the news brings home the importance of this chapter. In January 1994, following heavy rains Europe's rivers were at the limit when *jahrhunderthochwasser* (high water not seen for 100 years). The Rhine River at Bonn rose 10m above flood stage, highest since the 1700s. France, Netherlands, and Italy also experienced flooding. October 1994 found southwest Texas under a downpour of over 50 cm in less than two days. An unusual consequence of the erosion of San Jacinto river channel by the heavy discharge was a break in a major gasoline pipeline—one that carries almost 20 percent of the U. S. supply. Flames exploded as 200,000 barrels of fuel shot to the surface. River flooding occurred throughout the region. The Piedmont region of northern Italy was hit with 25 cm of rain in November 1994, leaving 10,000 homeless, 65 dead, and $4 billion in lost property. Thus, we begin our important study of rivers.

Learning Objectives

The following learning objectives help guide your reading and comprehension efforts. The operative word is in *italics*. Read and work with these carefully and note that exercise #1 asks you about five of these objectives. After reading the chapter and using this workbook, you should be able to:

1. *Recall* stream discharge from Table 6-1, p. 203, and *list* Earth's major rivers.
2. *Define* the term fluvial and *outline* the fluvial processes: erosion, transportation, and deposition.
3. *Construct* a basic drainage basin model and *differentiate* between drainage divides and watersheds.
4. *List* the major drainage divides and drainage basins in the United States and Canada.
5. *Determine* in which region relative to the continental divides (Figure 11-1) and in

which river drainage basin (Figure 11-2) you live.

6. *Define* drainage patterns that occur in nature.
7. *Identify* the type of drainage pattern depicted in the satellite image in Figure 11-3, p. 346, and *explain* the structural basis for this pattern.
8. *Describe* the type of drainage pattern in the parallel-trending Valley and Ridge Province of the eastern United States (refer back to the topographic map/*Landsat* image composite featured in Figure 9-12, p. 295).
9. *Analyze* the behavior of a stream channel during a flood and *describe* the effects of increased discharge.
10. *Explain* the various ways that a stream transports its load and *identify* the relationship between velocity and sediment size to erosion, transport, and deposition.
11. *Explain* the effects on a stream channel of a flood discharge.
12. *Portray* a braided stream pattern and *describe* the conditions that produce such a pattern.
13. *Develop* a model of a meandering stream and *relate* the concepts of point bar, cut bank, and cutoff.
14. *Relate* the perils of basing a political boundary on a meandering stream channel.
15. *Interpret* the evolution of Niagara Falls and *relate* this to the concept of a nickpoint.
16. *Define* a floodplain and *list* related features and their interaction with the channel.
17. *Differentiate* the several types of river deltas and *detail* each.
18. *Explain* what is meant by a 10-year flood and *review* the fact that this does not affect the probability of an occurrence in any single year.
19. *Describe* the methods of measuring streamflow.
20. *Illustrate* with a simple sketch a stream hydrograph for a typical drainage basin and *characterize* it before and after urbanization.
21. *Define* a PMF and PMP as a planning basis for flood events.

Outline Headings and Glossary Review

These are the first- second-, and third-orde headings that divide this chapter. The ke terms and concepts that appear **boldface** in th text are listed under their appropriate headin in bold italics; these highlighted terms appea in the text glossary. A check-off box is place next to each key term so you can mark you progress through the chapter as you define thes in your reading notes or prepare note cards.

Fluvial Processes and Landscapes

- ❑ ***fluvial***
- ❑ ***erosion***
- ❑ ***transport***
- ❑ ***deposition***
- ❑ ***alluvium***

The Drainage Basin System

- ❑ ***drainage basin***
- ❑ ***watershed***
- ❑ ***continental divides***

Drainage Patterns

- ❑ ***sheet flow***
- ❑ ***drainage pattern***

Streamflow Characteristics

Stream Erosion

- ❑ ***hydraulic action***
- ❑ ***abrasion***

Stream Transport

- ❑ ***dissolved load***
- ❑ ***suspended load***
- ❑ ***bed load***
- ❑ ***traction***
- ❑ ***saltation***
- ❑ ***aggradation***
- ❑ ***braided stream***

Channel Patterns

- ❑ ***meandering stream***
- ❑ ***cut bank***
- ❑ ***point bar***
- ❑ ***oxbow lake***

Stream Gradient

- ❑ ***gradient***
- ❑ ***graded stream***
- ❑ ***entrenched meanders***

Nickpoints
- ❑ ***nickpoint***

Stream Deposition

Floodplains
- ❑ ***floodplain***
- ❑ ***natural levees***

Stream Terraces
- ❑ ***alluvial terraces***

River Deltas
- ❑ ***delta***
- ❑ ***estuary***

❑ Floods and River Management

Streamflow Measurement
- ❑ ***flood***
- ❑ ***hydrographs***

SUMMARY

Learning Activities and Critical Thinking

1. Select any five learning objectives from the list presented at the beginning of this chapter. Place the number selected in the space provided (no need to rewrite each objective). Using the following questions as guidelines only, briefly discuss your treatment of the objective.

- What did you know about the objective before you began?
- What was your plan to complete the objective?
- Which information source did you use in your learning (text, or other)?
- Were you able to complete the action stated in the objective? What did you learn?
- Are there any aspects of the objective about which you want to know more?

a)____:____

____.

b)____:____

____.

c)____:____

____

____

d)____:___________________________________

____

____

____

____

____

e)____:___________________________________

____

____

____

____

____

2. In terms of average discharge at their mouths, list the six greatest rivers on Earth, their discharge amounts, and the body of water into which they flow (see Table 6-1, p. 203).

(a) ______________________________________

(b) ______________________________________

(c) ______________________________________

(d) ______________________________________

(e) ______________________________________

(f) ______________________________________

3. Complete the following: "Wind, water, and ice dislodge, dissolve, or remove surface material in the process called ____________________. Thus, streams produce *fluvial erosion,* which supplies weathered sediment for ____to new locations, where it is laid down in a process known as ____________________.
A stream is a mixture of water and solids—carried in solution, suspension, and by mechanical transport. ____________________ is the general term for the clay, silt, and sand transported by running water."

4. List the five principal drainage basins, divided by continental divides, for the United States and Canada (Figure 11-1, p. 343).

(a)____

(b)____

(c)____

(d)____

(e)____

5. Using the following map of the United States and Canada, label the major river drainage basins as identified in Figure 11-2 (p. 344). Also add to your map color lines that designate the major continental divides that separate Pacific, Atlantic, Gulf of Mexico, and Hudson Bay/Arctic Ocean drainage (consult Figure 11-1, p. 343 for these divides).

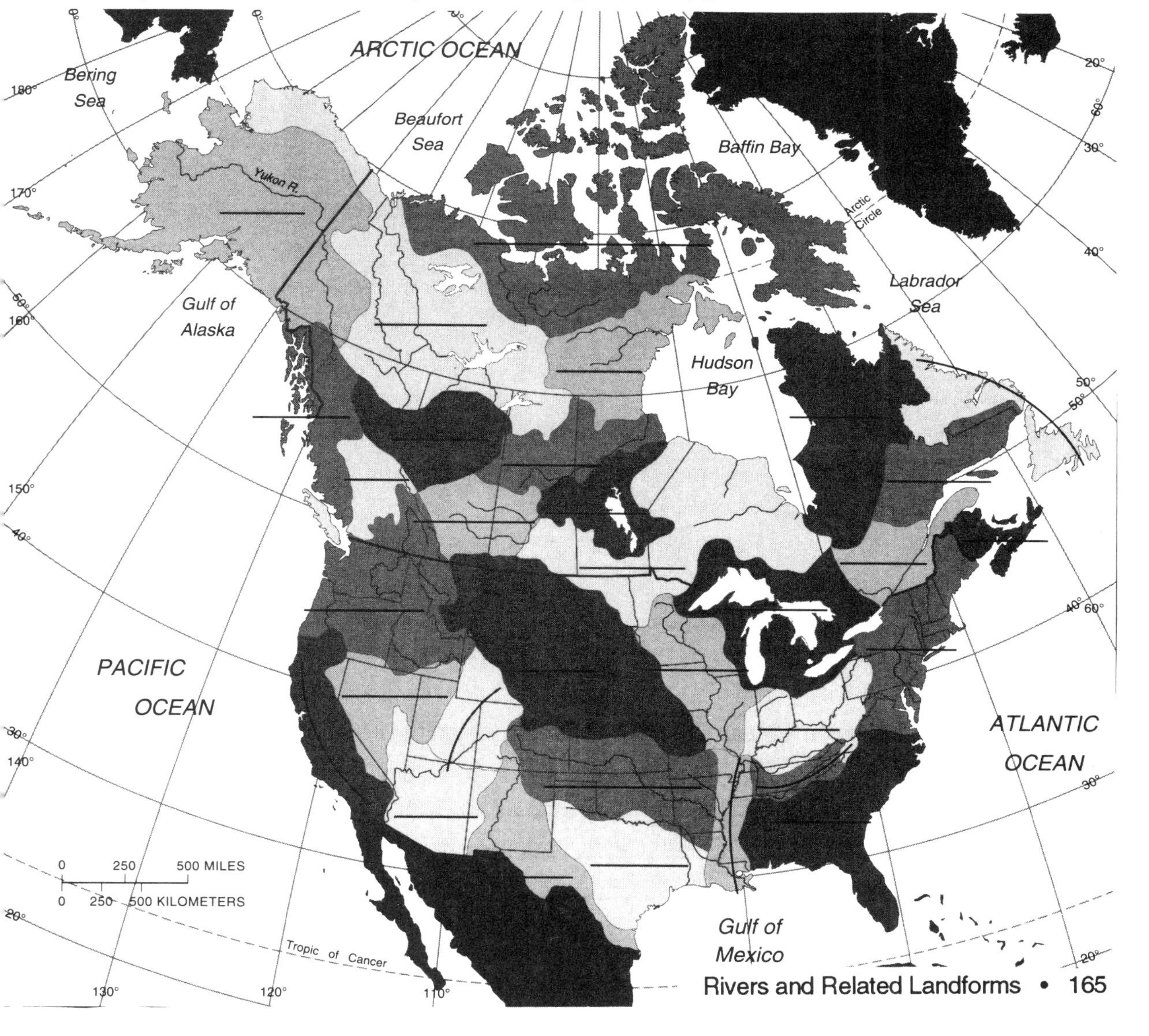

6. Identify and label each of the following drainage patterns (Figure 11-4, p. 347). Then mark on this illustration the appropriate figure number which is characteristic of: ❑ **(a)** the region shown in Figure 11-3, p. 346; ❑ **(b)** the southwest corner of "Topographic Map T.5–Jackson, Michigan" (presented at the end of the text after the Index); and ❑ **(c)** the Appalachian Mountain region shown by the image and topographic map in Figure 9-12, p. 295.

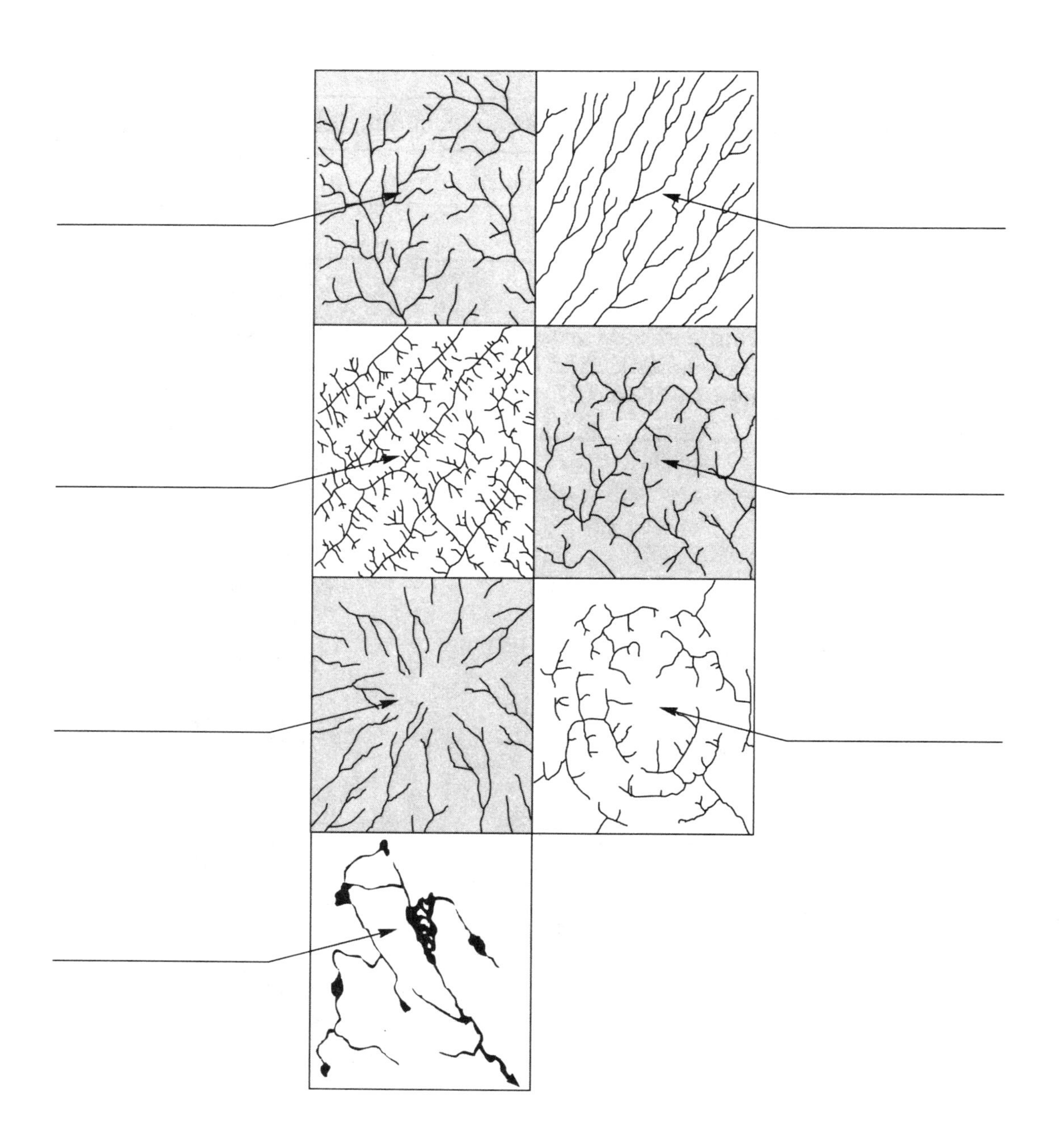

7. Differentiate between stream competence and capacity as they relate to stream transport ability.

(a) competence____

____

____.

(b) capacity____

____

____.

8. Relative to the flood stages of the San Juan River channel in Utah depicted in Figure 11-5, p. 348, what was occurring in terms of gauge height and discharge on each of the following dates:

(a) September 15, 1941____

____

____.

(b) October 14, 1941____

____

____.

(c) October 26, 1941____

____

____.

9. Analyze the actions and processes that are occurring in the two photos, Figure 11-6 a and b. Based on the text discussion, include information about the energy and work going on in each scene.

(a) Figure 11-6a:____

____

____

____.

(b) Figure 11-6 b:____

____

____

____.

10. In the space provided, diagram a stream cross section with an ideal *longitudinal profile* that illustrates a typical stream gradient. Add to the stream profile a *nickpoint* created by resistant rock strata (see Figures 11-11, p. 353, and 11-13, pp. 354).

11. Figure 11-10, p. 352, depicts the evolution of river meanders and presents an example of where a former river meander is still in use as a political border between Nebraska and Iowa. Please turn to "Topographic Map 2 (T.2): Omaha North, Nebraska–Iowa" that is included in the back of your text after the Index. Using this map and the discussion in the text answer the following. The legend for topographic map symbols is in Appendix B.

(a) What is the contour interval on this map segment?____________________________.

Given this value, determine the elevation of the surface of Carter Lake____________________.

(b) How wide is the neck of land belonging to Iowa? Measure along Abbott Drive.

____.

(c) Do you identify any industrial activity in Carter Lake? What are the circular symbols and the small lines around them?____

____.

(d) Explain the sequence of events that produced the present circumstances surrounding Carter Lake, Iowa. Consult the sequence shown in the Figure 11-10 and the discussion in the text as you prepare your answer.

____

____

____

____

____.

12. On the following illustration of a typical floodplain, derived from Figure 11-15, p. 355, label the components in the appropriate spaces noted.

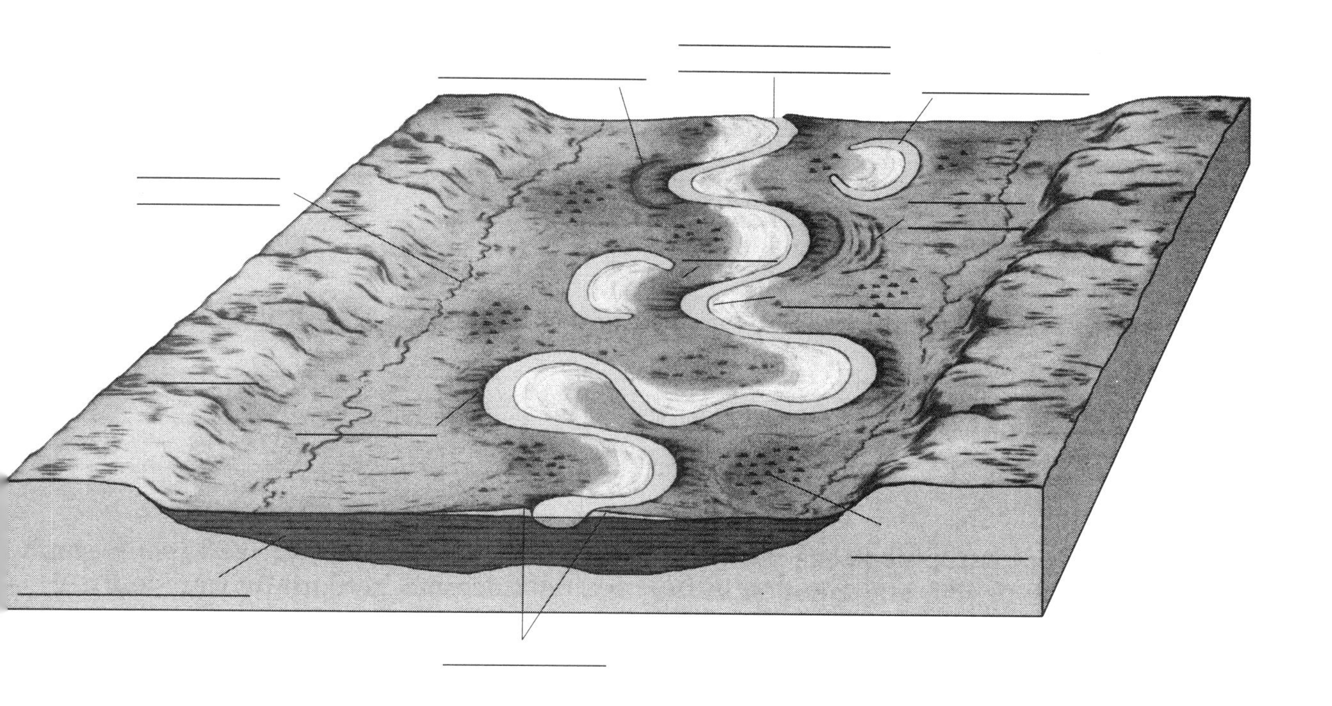

13. "Topographic Map 1 (T.1): Philipp, Mississippi" is included in the back of your text after the Index. The area shown on the map segment is a portion of the lower Mississippi River floodplain. Using this map and the discussion in the text answer the following. The legend for topographic map symbols is in Appendix B, p. B.2.

(a) What is the contour interval for this map segment?_____________.

(b) What is the spot highest elevation ("X") on the map?_____________. What is the elevation of the surface of the river?_________________.

(c) Are there levees along the river? If so, can you tell how high they are and how they are portrayed?____

____

____.

(d) What is the name of the water feature next to the Jones Chapel Cemetery?______

____.

How many of these water features can you identify on the topographic map?________

____.

14. Briefly overview the details of the 1993 Midwest flood (News Report #3, p. 358, and text). What lessons were learned concerning floodplains, zoning, levee management, etc.?

____

____

____

____

____

____.

15. Describe the relationship between a depositional floodplain and the development of stream terraces. How might stream terraces (paired and unpaired) help in the interpretation of the stream's development in a valley?

____

____

____

____.

16. After examining the Space Shuttle photograph in Figure 11-18, p. 358, explain the appearance of the Mouths of the Ganges and the condition of the Bay of Bengal at the mouths. Identify the ecological relationships among forestry, sediment, floodplain, and the conditions in Bangladesh and India (see text p. 361). You may want to refer back to News Report #1, p. 228, for more information.

____

____

____.

17. Briefly describe the evolution of the Mississippi River delta over the past 5000 years (Figure 11-20, p. 360). What is one probable fate of this region in the near- to middle-term future according to the text discussion?

____

____

____.

18. Describe what is meant by "Settlement Control Beats Flood Control" as discussed on page 365.____

____

____.

19. Label the following typical stream hydrograph for a drainage basin, taking care to identify those portions that apply to urbanization both before and after development (Figure 11-22, p. 366).

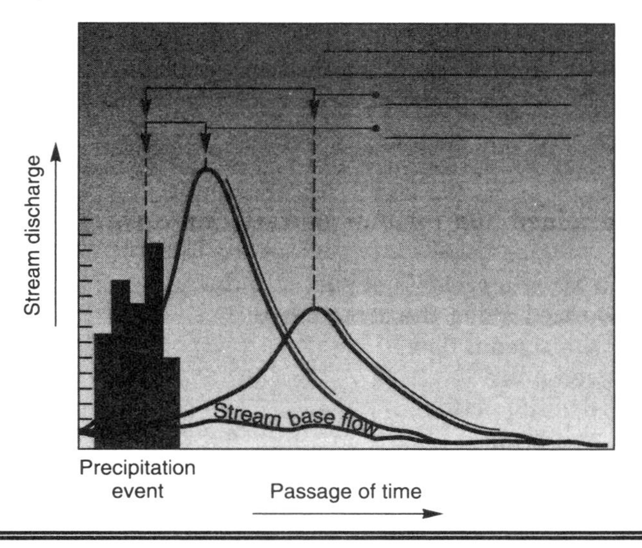

Sample Self-test

(Answers appear at the end of the study guide.)

1. Watersheds are defined by
   a) continental divides
   b) drainage divides
   c) unit hydrographs
   d) stream orders

2. Which of the following is <u>incorrectly</u> matched?
   a) rills – deep stream valleys
   b) gullies – developed rills
   c) drainage divides – ridges that control drainage
   d) Delaware River basin – in parts of five states

3. The Valley and Ridge Province is characterized by which drainage pattern?
   a) dendritic
   b) trellis
   c) parallel
   d) annular

4. The area depicted in Figure 11-3, p. 346, near the junction of the West Virginia, Ohio, and Kentucky borders is characterized by which drainage pattern?
   a) dendritic
   b) trellis
   c) parallel
   d) annular

5. The Amazon River has
   a) the greatest discharge of any river on Earth
   b) not produced a delta in the Atlantic Ocean
   c) a braided mouth with many islands
   d) all of these are correct

6. The suspended load of a stream consists of particles that
   a) are rolled and bounced along the stream bed
   b) are held aloft in the stream flow
   c) drag along the stream bed
   d) are basically in solution

7. The downstream portion of a river
   a) generally becomes more sluggish
   b) generally is of higher velocity, which is masked by reduced turbulence
   c) usually has turbulent flows
   d) has lower discharges than do upstream portions

8. Relative to a meandering stream
   a) the inner portion of a meander features a cut bank
   b) the inner portion of a meander features a point bar
   c) they tend to be straight
   d) they tend to develop when a steep gradient is formed

9. Niagara Falls is an example of a nickpoint related to differing resistances of bedrock strata.
   a) true
   b) false

10. Excess sediment in a stream will produce a maze of interconnected channels causing the stream to be braided.
   a) true
   b) false

11. Relative to stream gradient, it is not possible to have both ungraded and graded portions on the same stream.
   a) true
   b) false

12. The area of maximum velocity in a stream is usually in the inside of a bend near the point bar.
   a) true
   b) false

13. Natural levees are actually formed as by-products of flooding even though people depend on them for possible protection.
   a) true
   b) false

14. An oxbow lake is a form of braided stream.
   a) true
   b) false

15. The Nile River, which flows into the Mediterranean Sea, is an example of an arcuate delta.
   a) true
   b) false

16. Streamflows are measured with a staff gauge and stilling well, among other devices and techniques.
   a) true
   b) false

17. Urbanization both delays and lessens peak flow as plotted on a hydrograph.
   a) true
   b) false

18. The Mississippi-Missouri-Ohio river system is a single drainage basin rather than many smaller drainage basins considered separately.

a) true
b) false

19. Relative to continental divides, the Delaware River flows into the ____, the Mississippi River flows into the ____ the Columbia River flows into the ____, and the Yukon River flows into the ____.

20. Stream drainage patterns associated with steep slopes and some relief are usually____________ those with a faulted and jointed landscape are ________________; those with volcanic peaks are ____________________________; and those associated with structural domes and concentric rock patterns are ____.

21. The____ is now blocked off from the Mississippi River at the Old Control Structure, yet is one-half the distance of the present Mississippi River channel to the Gulf of Mexico and is the site for a potential future channel change.

22. During the past 5000 years, the Mississippi River delta has assumed ________________ (number of) deltaic forms. The present bird-foot delta has been building for the past ______________ years at least.

Name:______________________________ Class Section:________________

Date:__________________ Score/Grade:____________

Wind Processes and Desert Landscapes

12

Chapter Overview

Wind is an agent of erosion, transportation, and deposition. Its effectiveness has been the subject of much debate; in fact, wind at times was thought to produce major landforms. Presently, wind is regarded as a relatively minor exogenic agent, but it is significant enough to deserve our attention. In this chapter we examine the work of wind, its associated processes, and resulting landforms.

In addition, we discuss desert landscapes, where water remains the major erosional force but where an overall lack of moisture and stabilizing vegetation allows wind processes to operate. Evidence of this includes sand seas and sand dunes of infinite variety. Beach and coastal dunes, which form in many climates, also are influenced by wind and are discussed in this chapter.

Learning Objectives

The following learning objectives help guide your reading and comprehension efforts. The operative word is in *italics*. Read and work with these carefully and note that exercise #1 asks you about five of these objectives. After reading the chapter and using this workbook, you should be able to:

1. *Characterize* the unique aspects of the work of the wind and eolian processes.
2. *Outline* the work completed by Ralph Bagnold as described in the *Physics of Blown Sand and Desert Dunes*.
3. *Analyze* wind velocity as it relates to sand movement.
4. *Describe* eolian erosion and *explain* deflation, abrasion, and related resultant landforms.
5. *Interpret* the impact of the 1991 Persian Gulf War on desert pavement formations and *list* the related names for these desert pavements in other parts of the world.
6. *Define* a ventifact and *portray* at least two examples including a yardang.
7. *Describe* eolian transportation and *explain* saltation, surface creep, fluid threshold and impact threshold.
8. *Differentiate* between a reg desert and an erg desert and *relate* which is more dominant on Earth.
9. *Identify* the major classes of sand dunes and *recite* examples within each class.
10. *Explain* loess deposits and *identify* their origins in North America and Europe and in China.
11. *Portray* desert landscapes in terms of climatic factors, including temperature, energy balances, and moisture, and *locate* these regions on a world map.

12. *Identify* water as the major erosional force in the desert.
13. *Describe* desert fluvial processes and *relate* these to the landforms they create.
14. *Explain* the formation of desert alluvial fans and their coalesced form as a bajada.
15. *Interpret* the two photographs of the Stovepipe Wells dune field presented in Figure 12-12, p. 384.
16. *Overview* the Colorado River basin in both historical and physical terms and *characterize* its annual discharge as that of an exotic stream.
17. *Analyze* the water budget of the Colorado River and *explain* any discrepancies between supply and demand that you discern.
18. *Define* and *characterize* the Basin and Range Province of the western United States.
19. *Diagram* a bolson and *label* the various physical elements portrayed.
20. *Differentiate* between bajada and pediment.

Outline Headings and Glossary Review

These are the first- second-, and third-order headings that divide this chapter. The key terms and concepts that appear **boldface** in the text are listed under their appropriate heading in bold italics; these highlighted terms appear in the text glossary. A check-off box is placed next to each key term so you can mark your progress through the chapter as you define thes in your reading notes or prepare note cards.

The Work of the Wind

❑ ***eolian***

Eolian Erosion

❑ ***deflation***
❑ ***abrasion***

Deflation

❑ ***desert pavement***
❑ ***blowout depression***

Abrasion

❑ ***ventifacts***

Eolian Transportation

❑ ***surface creep***

Eolian Depositional Landforms

❑ ***dune***
❑ ***erg desert***
❑ ***sand sea***

Dune Movement and Form

❑ ***slipface***

Loess Deposits

❑ ***loess***

Overview of Desert Landscapes

Desert Climates

Desert Fluvial Processes

❑ ***flash flood***
❑ ***wash***
❑ ***playa***

Alluvial Fans

❑ ***alluvial fan***
❑ ***bajada***

Desert Landscapes

❑ ***badland***

Basin and Range Province

❑ ***Basin and Range Province***
❑ ***bolson***

SUMMARY

Learning Activities and Critical Thinking

1. Select any five learning objectives from the list presented at the beginning of this chapter. Place the number selected in the space provided (no need to rewrite each objective). Using the following questions as guidelines only, briefly discuss your treatmen of the objective.

- What did you know about the objective before you began?
- What was your plan to complete the objective?
- Which information source did you use in your learning (text, or other)?

- Were you able to complete the action stated in the objective? What did you learn?
- Are there any aspects of the objective about which you want to know more?

a)____:____

____

____

____

____

____.

b)____:____

____

____

____

____

____.

c)____:____

____

____

____

____

____.

d)____:____

____

____

____

____

____.

e)____:____

____

____

____

____

____.

2. Describe the relationship between sand movement and wind velocity as measured over a meter-wide strip of ground surface. ______________________________

____

____

______________________________.

3. An extensive area of sand and sand dunes is called an ____________ desert or a ____________. In contrast most of Earth's deserts comprise desert pavement; some provincial names for these include (give at least three with locations):

a)______________________________

b)______________________________

c)______________________________

4. Briefly explain the relationship between the Persian Gulf War (1991) and deflation rates in the region (News Report #1, p. 373). ______________________________

____

____

______________________________.

5. In terms of eolian erosion and transport, what is shown in the photograph in Figure 12-3, p. 376, and supported in the text, that relates to human impact and intervention with coastal dunes?

____

____

______________________________.

6. Briefly compare and contrast *eolian transportation* and *fluvial transportation* (pp. 374-75 and pp. 348-50).

____

____

____

______________________________.

7. Label the cross-section illustration of a sand dune that appears in Figure 12-6 p. 377.

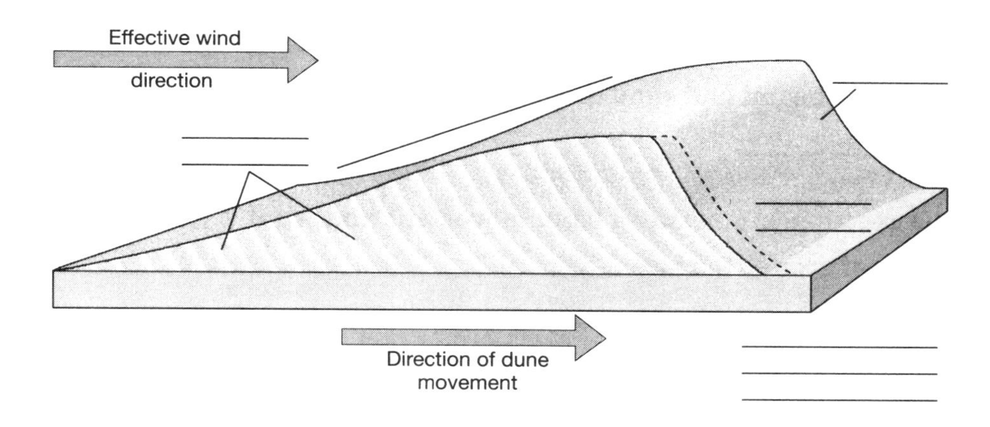

8. Define *loess* and describe both the glacial and the desert sources and processes that produce these materials. Refer to the photographs in Figure 12-9a and b, p. 381________

____

____

____

____.

9. Although rare, a precipitation event in the desert is often dramatic. Describe what occurs following a downpour in the desert (examine Figure 12-12, p. 384).____________

____

____

____

____

____

____.

10. Identify each type of sand dune pictured below as derived from Figure 12-7, p. 378-89, and note the effective or prevailing wind direction involved in the formation of each.

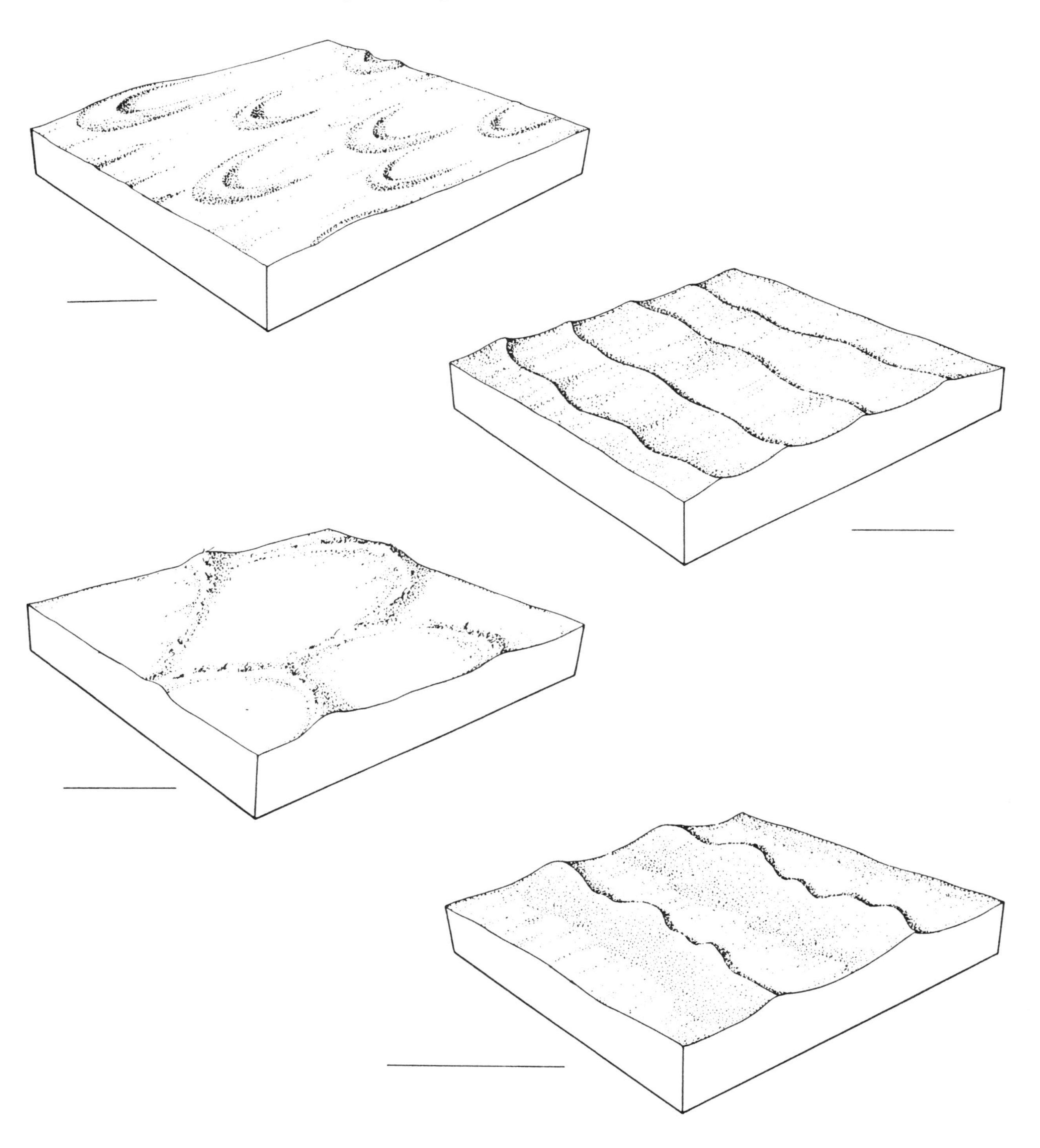

11. Using the arid-lands map in Figure 12-11, p. 383, and the text explanation, describe the processes that produce this spatial distribution of deserts and steppes—detail at least three locations on Earth and related processes.

(a)____

____;

(b)____

____;

(c)____

____.

12. Construct a simple bar graph depicting the Colorado River discharge at Lees Ferry for the years 1917, 1934, 1977, and 1984, and the 1930 through 1980 average flow of the river (see FYI Report 12-1, pp. 386-89 in the text and Table 1).

Use the years noted along the bottom axis to begin the five bar graphs. On the side axis, design and add an appropriate scale of flow-discharge amounts (in acre-feet per year) to use in plotting the height of each bar graph.

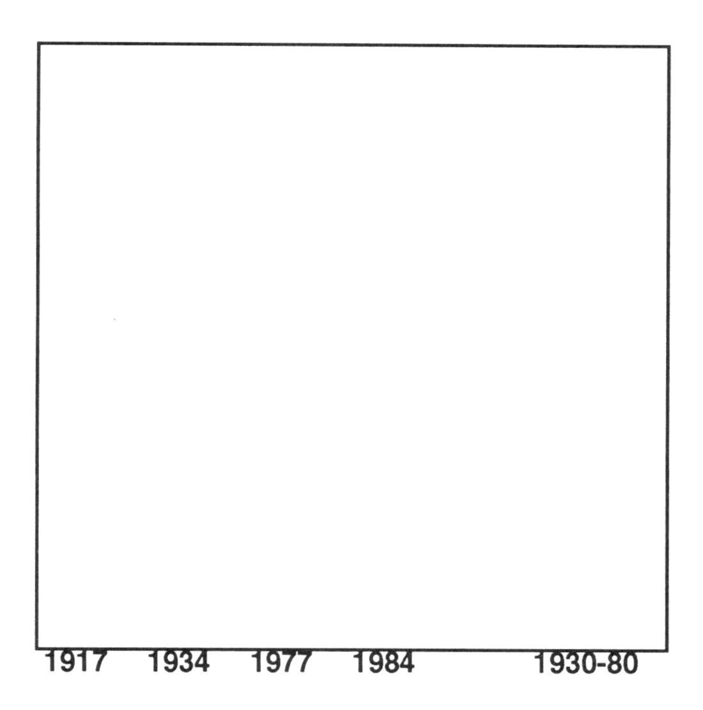

13. What is your assessment of the data presented in Table 1, p. 388?________________

____

____

____

____.

14. According to the FYI Report, what was the impact of Glen Canyon Dam near Lees Ferry on the overall discharge of the Colorado River? Be specific.

____

____

____

____.

15. Figure 12-13, p. 385, presents a photograph and a topographic map of an alluvial fan—a typical landform feature in a desert environment. (The legend for topographic ma symbols is in Appendix B, p. B.2.) Please complete the following questions.

(a) What is the contour interval on this map segment?________________________

(b) What is the elevation at Lawton Ranch, near the mouth of the canyon?____________; and, what is the elevation at the village of Jeffers in the northwest portion of the map segment?________________. Given these two elevation values, what is the local relief of the alluvial fan and valley?__________________________.

(c) What is the highest elevation shown in the mountain range along the eastern margin of the map?______________. What is the local relief between this point and Jeffers? ________________.

(d) What processes produce an alluvial fan? Why do these generally not occur in humid regions?

____

____

____

____.

16. Prepare a brief list of some of Earth's great deserts, name and location (text pp. 383-84):

____

____

____

____.

17. Complete the labels and leader lines for Figure 12-17, p. 392, that portray a *bolson* in a desert landscape.

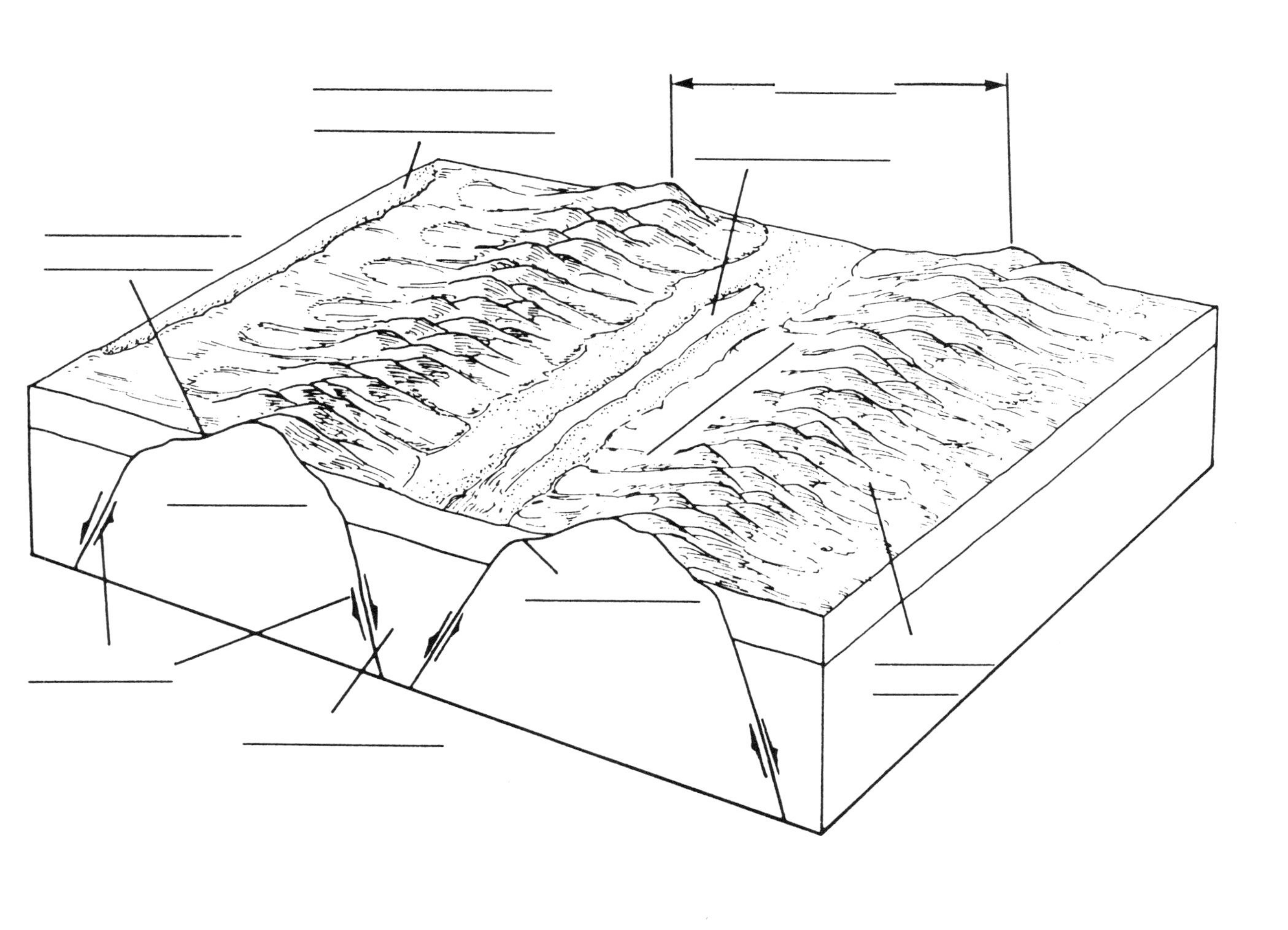

Sample Self-test

(Answers appear at the end of the study guide.)

1. Dry, arid climates occupy approximately what percentage of Earth's land surface?
   - a) 55 percent
   - b) 44 percent
   - c) 26 percent, and 35% including all semi-arid lands
   - d) 12 percent

2. The term eolian refers to
   - a) stream-related processes
   - b) sand dune fields only
   - c) wind-eroded, wind-transported, and wind-deposited
   - d) weathering and mass movement in humid regions

3. The formation of desert pavement
   - a) involves both wind and water
   - b) is produced by abrasion only
   - c) is not related to deflation
   - d) is greatly restricted in its occurrence on Earth

4. Nature's version of sandblasting is called
   - a) deflation
   - b) eolian deposition
   - c) abrasion
   - d) saltation

5. The Grand Ar Rub' al Khali is an example of a
   - a) gibber plain
   - b) reg desert
   - c) sand sea
   - d) stream-eroded area in the Middle East

6. Relative to loess deposits,
   - a) they are principally composed of sands and gravels
   - b) they are formed of fine-grained clays and silts
   - c) occurrences are found only in the United States
   - d) they form only following glacial activity

7. Which of the following correctly describes a dry stream bed that is intermittently filled with water?
   - a) wash
   - b) wadi
   - c) arroyo
   - d) all of these are correct
   - e) none of these is correct

8. An exotic stream
   a) is any river that flows from an arid region to one of adequate precipitation
   b) is one involved in a particular historical event
   c) is one whose entire course is in a desert region
   d) is any river that arises in a humid region and then passes through to an arid region

9. Relative to the discharge of the Colorado River system, the construction of a large multi-purpose dam such as Glen Canyon augments (increases) stream flows and helps sustain the available water in the river.
   a) true
   b) false

10. The flood control measures and facilities on the Colorado River worked to prevent flooding during the spring of 1983.
   a) true
   b) false

11. The Qattara Depression in Egypt is a product of deflation.
   a) true
   b) false

12. Loess deposits form loose, gradual slopes, unstable for construction or road cuts.
   a) true
   b) false

13. The smallest features produced by saltation are called ripples.
   a) true
   b) false

14. Loess deposits in China exceed 300 m (984 ft).
   a) true
   b) false

15. Desert regions are characterized by high potential evapotranspiration, low precipitation, high evaporation, high input of insolation, and high radiative heat losses at night.
   a) true b) false

16. Some desert plants depend on the rushing, crashing waters in a wash to germinate.
   a) true b) false

17. John Wesley Powell specifically called on the federal government to build large multi-purpose dams and reservoirs in the arid west.
   a) true b) false

18. The Basin and Range Province of the western United States is an example of a horst and graben landscape.
   a) true b) false

Name:________________________________ Class Section:__________________

Date:____________________ Score/Grade:______________

Coastal Processes and Landforms

13

Chapter Overview

Coastal regions are unique and dynamic environments. Most of Earth's coastlines are relatively new and are the setting for continuous change. The land, ocean, and atmosphere interact to produce tides, waves, erosional features, and depositional features along the continental margins. The interaction of vast oceanic and atmospheric masses is dramatic along a shoreline. At times, the ocean attacks the coast in a stormy rage of erosive power; at other times, the moist sea breeze, salty mist, and repetitive motion of the water are gentle and calming.

As you read Chapter 13 and learn of the coastal environment, remember the role of humans and human impacts on the coastline. The chapter builds to a discussion of certain human impacts and a focus study discussing an environmental approach to shoreline planning.

Learning Objectives

The following learning objectives help guide your reading and comprehension efforts. The operative word is in *italics*. Read and work with these carefully and note that exercise #1 asks you about five of these objectives. After reading the chapter and using this workbook, you should be able to:

1. *Identify* the components of the coastal environment and *label* each on a diagram.
2. *Define* mean sea level and *portray* the variability of average sea level along the coastline of the United States.
3. *List* the environmental inputs to the coastal system.
4. *Analyze* the occurrence of tides on Earth and *explain* the factors that create tidal phenomena.
5. *Differentiate* between spring tides and neap tides and *explain* the astronomical relationships that cause them.
6. *Relate* the development of electrical generation that utilizes tidal power.
7. *Portray* wave motion at sea and near shore and *differentiate* between waves of transition and waves of translation.
8. *Interpret* coastal straightening as a product of wave refraction.
9. *Describe* longshore current and littoral and beach drift.
10. *Portray* a tsunami, or seismic sea wave and *relate* its probable causes.
11. *Analyze* and *describe* any changes that have occurred, and may occur, in sea level worldwide and *identify* trends.
12. *Discern* the geomorphic difference between the Pacific and the Atlantic coastlines of

the United States and *explain* this difference.

13. *Identify* and *label* characteristic coastal erosional landforms.
14. *Identify* and *label* characteristic coastal depositional landforms.
15. *Analyze* the dynamic relationships present on a beach.
16. *Describe* various strategies of human intervention utilized to interrupt littoral and beach drift and longshore currents.
17. *Relate* present thinking on the formation of offshore barrier chains.
18. *Describe* barrier island hazards and *portray* human hazard perception or lack of perception in these environments.
19. *Describe* coral formations and *locate* their distribution on a world map.
20. *Differentiate* geographically and descriptively between salt marshes and mangrove swamps.
21. *Construct* an environmentally sensitive model for settlement and land use along the New Jersey shore and *propose* proper zoning strategies.

Outline Headings and Glossary Review

These are the first- second-, and third-order headings that divide this chapter. The key terms and concepts that appear **boldface** in the text are listed under their appropriate heading in bold italics; these highlighted terms appear in the text glossary. A check-off box is placed next to each key term so you can mark your progress through the chapter as you define these in your reading notes or prepare note cards.

Coastal System Components

The Coastal Environment and Sea Level

- ❑ ***littoral zone***
- ❑ ***mean sea level***

Coastal System Actions

Tides

- ❑ ***tides***

Tidal Power

Waves

- ❑ ***waves***
- ❑ ***swells***
- ❑ ***breaker***

Wave Refraction

- ❑ ***headlands***
- ❑ ***wave refraction***
- ❑ ***longshore current***
- ❑ ***beach drift***

Tsunami, or Seismic Sea Wave

- ❑ ***tsunami***

Sea Level Changes

Coastal System Outputs

Erosional Coastal Processes and Landforms

- ❑ ***wave-cut platform***

Depositional Coastal Processes and Landforms

- ❑ ***barrier spit***
- ❑ ***bay barrier***
- ❑ ***lagoon***

Beaches

- ❑ ***beach***

Barrier Forms

- ❑ ***barrier beaches***
- ❑ ***barrier islands***

Emergent and Submergent Coastlines

Emergent Coastlines

Submergent Coastlines

Organic Processes: Coral Formations

- ❑ ***coral***

Coral Reefs

Salt Marshes and Mangrove Swamps

- ❑ ***salt marshes***
- ❑ ***mangrove swamps***

Human Impact on Coastal Environments

SUMMARY

Learning Activities and Critical Thinking

1. Select any five learning objectives from the list presented at the beginning of this chapter. Place the number selected in the space provided (no need to rewrite each objective). Using the following questions as guidelines only, briefly discuss your treatment of the objective.

- What did you know about the objective before you began?
- What was your plan to complete the objective?
- Which information source did you use in your learning (text, or other)?
- Were you able to complete the action stated in the objective? What did you learn?
- Are there any aspects of the objective about which you want to know more?

a)____:____

____

____

____

____

____.

b)____:____

____

____

____

____

____.

c)____:____

____

____

____

____

____.

d)____:____

____

____

____

____

____.

e)____:____

____

____

____

____

____.

2. Itemize the physical components (inputs) to the coastal system as discussed in the text

____

____

____

____.

3. Using the following illustration derived from Figure 13-1, p. 397, label the components noted for the littoral zone.

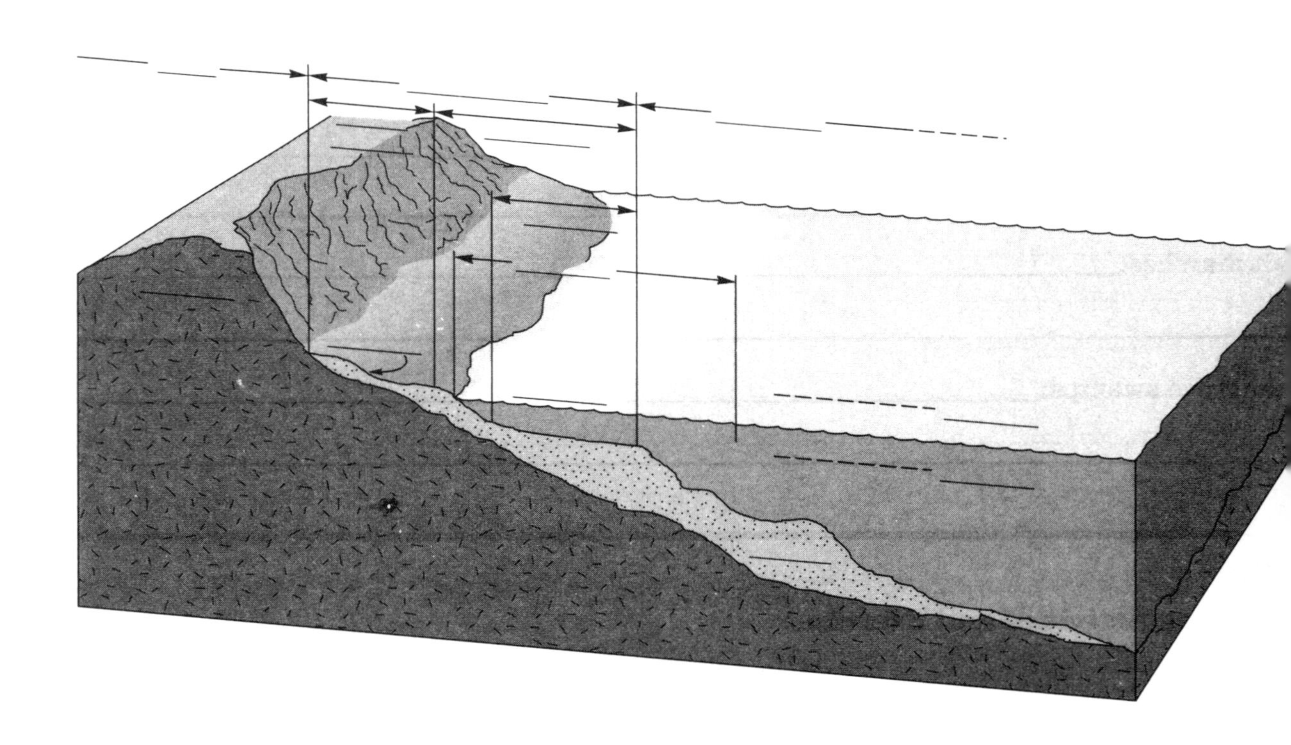

4. Describe the astronomical relationships that occur during the following times.

(a) spring tides:____.

(b) neap tides:____.

5. What is mean sea level?____.

(a) How does sea level vary along the U. S. Coastline (News Report #1, p. 398)?____.

(b) According to the text (pp. 404-05) what changes have occurred in sea level?

- 18,000 B.P.:____.
- 100 years ago:____

(c) Briefly describe the forecast for future sea levels.

- If Greenland and Antarctica's ice sheets were completely melted: ____.
- Worst-case sea-level rise in the next 100 years:____.
- Most probable sea-level rise in the next 100 years:____.

(d) How do ocean temperatures relate to sea level values (News Report #3, p. 405)? Relate the ocean temperature measurements gathered off of southern California's coastline. Explain.____.

6. Are there any tidal electrical generating stations operating presently in North America? If so, briefly describe it and explain the principle involved.

____

____

____.

7. Differentiate between *waves of transition* and *waves of translation*, include the physical conditions that are involved in each phase of wave action.

(a) transition:____

____

____.

(b) translation:____

____

____.

8. Complete the appropriate labeling for an erosional coastline using this illustration derived from Figure 13-7, p. 406.

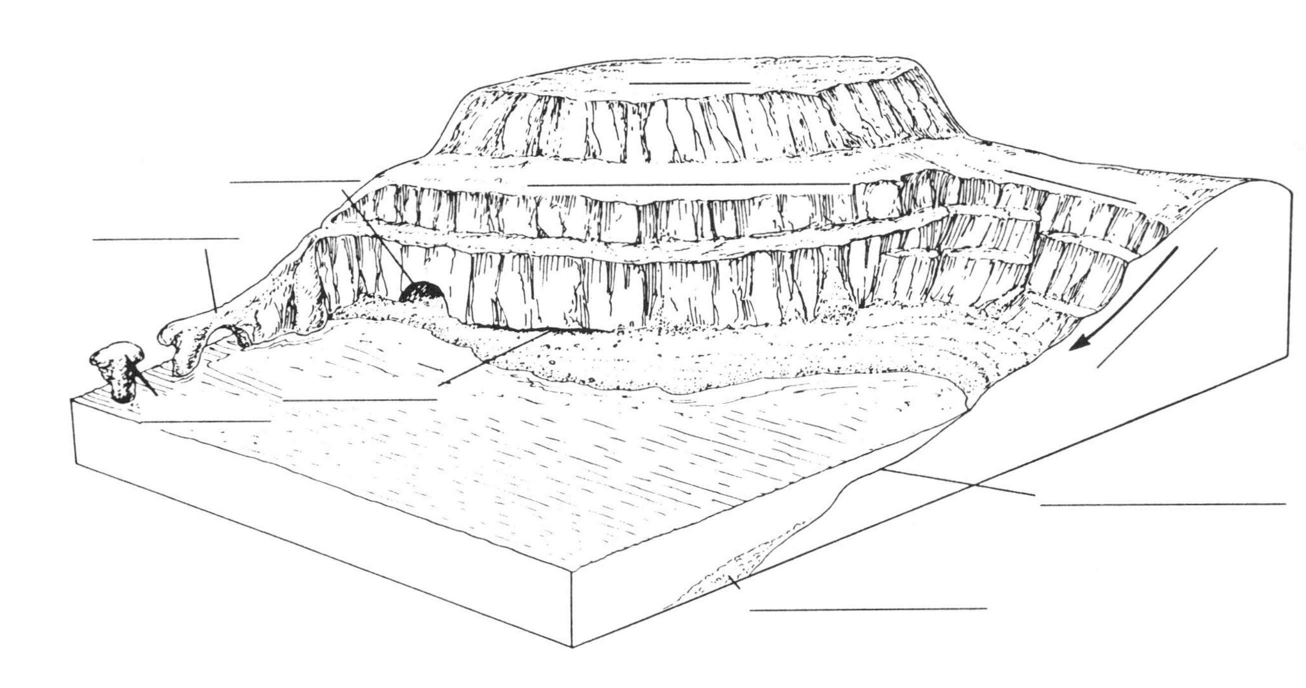

(a) Where is this type of erosional coastline found on Earth? Give a few examples beginning with the photo insert in Figure 13-7.

____

____.

(b) Describe what produced the landform pictured in Figure 13-8, p. 407.________________

____.

9. Complete the appropriate labeling for this depositional coastline derived from Figure 13-9, p. 407.

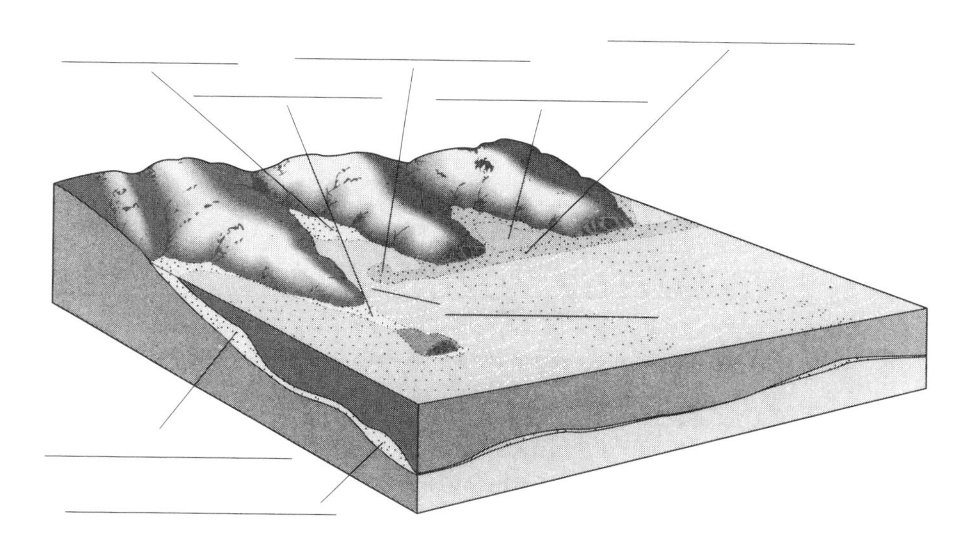

(a) Where is this type of depositional coastline found on Earth? Give a few examples beginning with the photo insert in Figure 13-9.

____

____.

10. Societies use several specific physical constructions to intervene in littoral and beach drift and longshore, or littoral, currents. Explain the function of each and mention any consequences that might result from each construction.

(a) breakwater:____

____

(b) jetty:____

____

(c) groin:____

____

11. Although not entirely satisfactory, one type of classification for coastlines is whether they appear to be *emergent* or *submergent*. Noting that some coastlines bear the traits of both, please characterize these two broad classifications (text pp. 409-10).

a) emergent:____

____

____

b) submergent:_______________________________________

____

____

12. "Topographic Map 6 (T.6): Point Reyes, California" is included in the back of your text after the Index. The area shown on the map segment is a portion of the central California coast just north of San Francisco. Using this map and the discussion in the text, answer the following. The legend for topographic map symbols is in Appendix B, p. B.2.

(a) What is the contour interval on this map?______________________

(b) Do you identify a barrier spit on the map segment? Name and location:________

____

(c) What is the composition of this barrier spit, according to the map?__________

(d) Can you determine the prevailing, effective wind direction along Point Reyes beach? Explain.

____

____

13. How is it possible for engineers to nourish(!) a beach (News Report #4)?________________

14. Relative to Earth's coral formations, what condition is emerging (News Report #6, p. 414)? Any clues as to why this is occurring?________________

15. Relative to the following illustration derived from FYI Report 13-1, Figure 1, p. 416, label and identify each of the components characteristic of the coastal environment along the New Jersey shore. Use the spaces provided for your labels and descriptions.

Sample Self-test

(Answers appear at the end of the study guide.

1. Another general term for the coastal environment is
   a) shoreline
   b) coast
   c) coastline
   d) littoral zone

2. Relative to mean sea level,
   a) a consistent value has yet to be determined due to all the variables involved in producing the tides
   b) it is at a similar level along all of the North American coast
   c) it is calculated based on average tidal levels recorded hourly at a given site over a period of 19 years
   d) sea level along the Gulf Coast is the lowest for the coasts of the lower 48 states

3. Which of the following is <u>incorrect</u>?
   a) rising tides – flood tides
   b) ebb tides – falling tides
   c) spring tides – happen only in the spring
   d) spring tides – extremes of high and low tides

4. Waves travel in wave trains that
   a) are produced by storm centers and generating regions far distant from where they break
   b) usually form relatively close to the affected coastline
   c) are developed in relation to the specific contours and shape of the affected coastline
   d) are called waves of translation as they travel in the open sea

5. A tsunami is
   a) also known as a tidal wave
   b) also known as a seismic sea wave
   c) a large wave form that travels in the open sea roughly at the same velocity that is witnessed at the shore
   d) caused by a strong storm center in the open sea

6. Sand deposited in a long ridge extending out from a coast that is connected to land at one end is called a
   a) barrier spit
   b) lagoon
   c) bay barrier
   d) tombolo

7. The most extensive chain of barrier islands in the world is along
   a) the western coast of Australia
   b) the coasts that surround the periphery of the Indian Ocean
   c) the east coast of the Asian landmass
   d) the Atlantic and Gulf coasts of North America, extending some 5000 km

8. Embayments and ria coasts are characteristic of
   a) erosional coasts
   b) emergent coasts
   c) submergent coasts
   d) the western margin of North America

9. The coastal environment is generally called the littoral zone.
   a) true
   b) false

10. Earth's tidal bulges are only produced by Earth's centrifugal forces as opposed to gravitational forces.
   a) true
   b) false

11. Earth's tidal bulges are about half attributable to the pull of the Sun, even though the Sun is many times the size of the Moon.
   a) true
   b) false

12. Spring tides reflect greater tidal ranges, whereas neap tides produce lesser tidal ranges.
   a) true
   b) false

13. Electricity has yet to be produced by tidal power generation but remains a viable alternative for the future.
   a) true
   b) false

14. A wave-cut platform, or wave-cut terrace, is an example of a coastal depositional feature.
   a) true
   b) false

15. Beaches are dominated by sand because quartz (SiO_2) is so abundant and resists weathering.
   a) true
   b) false

16. The barrier islands off South Carolina were struck for the <u>first</u> time by Hurricane Hugo in 1989, with the local residents never having experienced such storms before.
   a) true
   b) false

17. Corals are not capable of building major landforms; rather, they form on the top of already existing landmasses.
   a) true   b) false

18. A salt marsh is not a wetland, whereas a mangrove swamp is a wetland.
   a) true   b) false

19. Relative to possible development, the front, or primary dune, is________________________________
____; the trough behind the primary dune is__________
____; the secondary
dune is____; the backdune
is____; and the bayshore and
bay is____
____.

20. The circular undulations in the open sea are known as waves of __________________________,
whereas the wave action near shore forms breakers and waves of __________________________.

21. A__________________________consists of sand deposited in a long ridge extending out from a
coast; it partially crosses and blocks the mouth of a bay. This spit becomes a ___________________
_______________if it completely cuts off the bay from the ocean. A ________________________occurs when
sand deposits connect the shoreline with an offshore island or sea stack.

22. Relative to coastal wetlands, _______________________________ tend to form north of the 30th parallel
whereas______________________________________ form equatorward of that point.

Name:________________________ Class Section:________________

Date:______________ Score/Grade:____________

Glacial and Periglacial Landscapes

14

Chapter Overview

As the text states a large measure of the freshwater on Earth is frozen, with the bulk of that ice sitting restlessly in just two places—Greenland and Antarctica. The remaining ice covers various mountains and fills alpine valleys. More than 29 million km^3 (7 million mi^3) of water is tied up as ice, or about 77% of all freshwater. Distinctive landscapes are produced by the ebb and flow of glacial ice masses. Other regions experience permanently or seasonally frozen ground that effects the topography and human habitation.

These deposits of ice, laid down over several million years, provide an extensive frozen record of Earth's climatic history and perhaps some clues to its climatic future. The inference is that rather than being static and motionless, distant, frozen places, these frozen lands are dynamic and susceptible to change, just as they have been over Earth's history. Changes in the ice mass worldwide signal vast climatic change and further glacio-eustatic increases in sea level. As you study this chapter, keep this significance in mind.

Learning Objectives

The following learning objectives help guide your reading and comprehension efforts. The operative word is in *italics*. Read and work with these carefully and note that exercise #1 asks you about five of these objectives. After reading the chapter and using this workbook, you should be able to:

1. *Recite* the amount of water that is tied up in ice at the present time and *review* the amount of Earth's area so affected.
2. *Define* what a glacier is and *relate* the different general forms on land and sea.
3. *Differentiate* between alpine and continental glaciers and *list* the types of glaciers that fall within each classification.
4. *Analyze* the satellite image and high-altitude photograph in Figure 14-1 and *locate* the various glacial forms.
5. *Plot* a graph that demonstrates a glacial mass balance and *identify* an equilibrium line.
6. *Describe* the process involved in the formation of glacial ice.
7. *Portray* the mechanics of glacial movement.
8. *Identify* the phenomenon of glacial surges and *relate* some recent examples of this action.
9. *Explain* the processes of glacial erosion.
10. *Describe* characteristic erosional landforms associated with alpine glaciation.

11. *Describe* characteristic depositional landforms associated with alpine glaciation.
12. *Describe* characteristic erosional and depositional features associated with continental glaciation.
13. *Define* the term periglacial and *relate* the distribution of permafrost in the Northern Hemisphere.
14. *Describe* ground ice and frost action processes.
15. *Review* several of the human adaptations to living in permafrost and frozen ground environments.
16. *Explain* the Pleistocene ice age epoch and *relate* the timing and nature of both glacials and interglacials.
17. *Explain* present thinking as to the various mechanisms that bring on an ice age as described in FYI Report 14-1.
18. *Interpret* the occurrence of pluvial lakes and *relate* the modern dynamics of the Great Salt Lake.
19. *Describe* the nature of the Arctic and Antarctic regions.
20. *Portray* the massive Antarctic ice sheet and specifically *describe* the West Antarctic ice sheets.

Outline Headings and Glossary Review

These are the first- second-, and third-order headings that divide this chapter. The key terms and concepts that appear **boldface** in the text are listed under their appropriate heading in bold italics; these highlighted terms appear in the text glossary. A check-off box is placed next to each key term so you can mark your progress through the chapter as you define these in your reading notes or prepare note cards.

Reservoirs of Ice

❑ ***glacier***

Types of Glaciers

Alpine Glaciers

❑ ***alpine glacier***
❑ ***valley glacier***
❑ ***cirque***
❑ ***icebergs***

Continental Glaciers

❑ ***continental glacier***
❑ ***ice sheet***
❑ ***ice cap***
❑ ***ice field***

Glacial Processes

Formation of Glacial Ice

❑ ***firn***
❑ ***glacial ice***

Glacial Mass Balance

❑ ***firn line***
❑ ***ablation***
❑ ***equilibrium line***

Glacial Movement

❑ ***crevasses***

Glacial Surges

❑ ***glacial surge***

Glacial Erosion

❑ ***abrasion***

Glacial Landforms

Erosional Landforms Created by Alpine Glaciation

❑ ***arete***
❑ ***col***
❑ ***horn***
❑ ***fjord***

Depositional Landforms Created by Alpine Glaciation

❑ ***glacial drift***
❑ ***till***
❑ ***stratified drift***
❑ ***moraine***
❑ ***lateral moraine***
❑ ***medial moraine***

Erosional and Depositional Features of Continental Glaciation

❑ ***till plain***
❑ ***outwash plains***
❑ ***esker***
❑ ***kettle***
❑ ***kame***
❑ ***drumlin***

Periglacial Landscapes

❑ ***periglacial***

Permafrost

❑ ***permafrost***

Frozen Ground Phenomena

❑ ***ground ice***

Frost Action Processes

❑ ***ice wedge***

❑ *patterned ground*

Hillslope Processes

❑ *solifluction*

Humans and Periglacial Landscapes

The Pleistocene Ice Age Epoch

❑ *ice age*

Pluvial Periods and Paleolakes

❑ *paleolakes*

❑ *pluvial*

❑ *lacustrine deposits*

Arctic and Antarctic Regions

The Antarctic Ice Sheet

SUMMARY

Learning Activities and Critical Thinking

1. Select any five learning objectives from the list presented at the beginning of this chapter. Place the number selected in the space provided (no need to rewrite each objective). Using the following questions as guidelines only, briefly discuss your treatment of the objective.

- What did you know about the objective before you began?
- What was your plan to complete the objective?
- Which information source did you use in your learning (text, or other)?
- Were you able to complete the action stated in the objective? What did you learn?
- Are there any aspects of the objective about which you want to know more?

a)_____:____

____

____

____

____

____.

b)_____:____

____

____

____

____

____.

c)_____:____

____

____

____

____

____.

d)____ :____

____

____

____

____

____.

e)____ :____

____

____

____

____

____.

2. In the *Landsat* image (and inset map) of the Mount McKinley region shown in Figure 14-1a, p. 423, how many alpine glaciers or valley glaciers can you identify?

______________________________.

3. Name a specific example of each of the following as presented in the text and Figure 14-2:

a) alpine glacier:____.

b) ice field:____.

c) ice cap:____.

d) ice sheet (name 2):____.

4. Differentiate between a continental glacier and an alpine glacier.______________

____

____.

5. Briefly relate the process that produces glacial ice from snow.

____

____

____

____.

6. Complete the labels on the following illustration of a typical retreating alpine glacier derived from Figure 13-3b, p. 426.

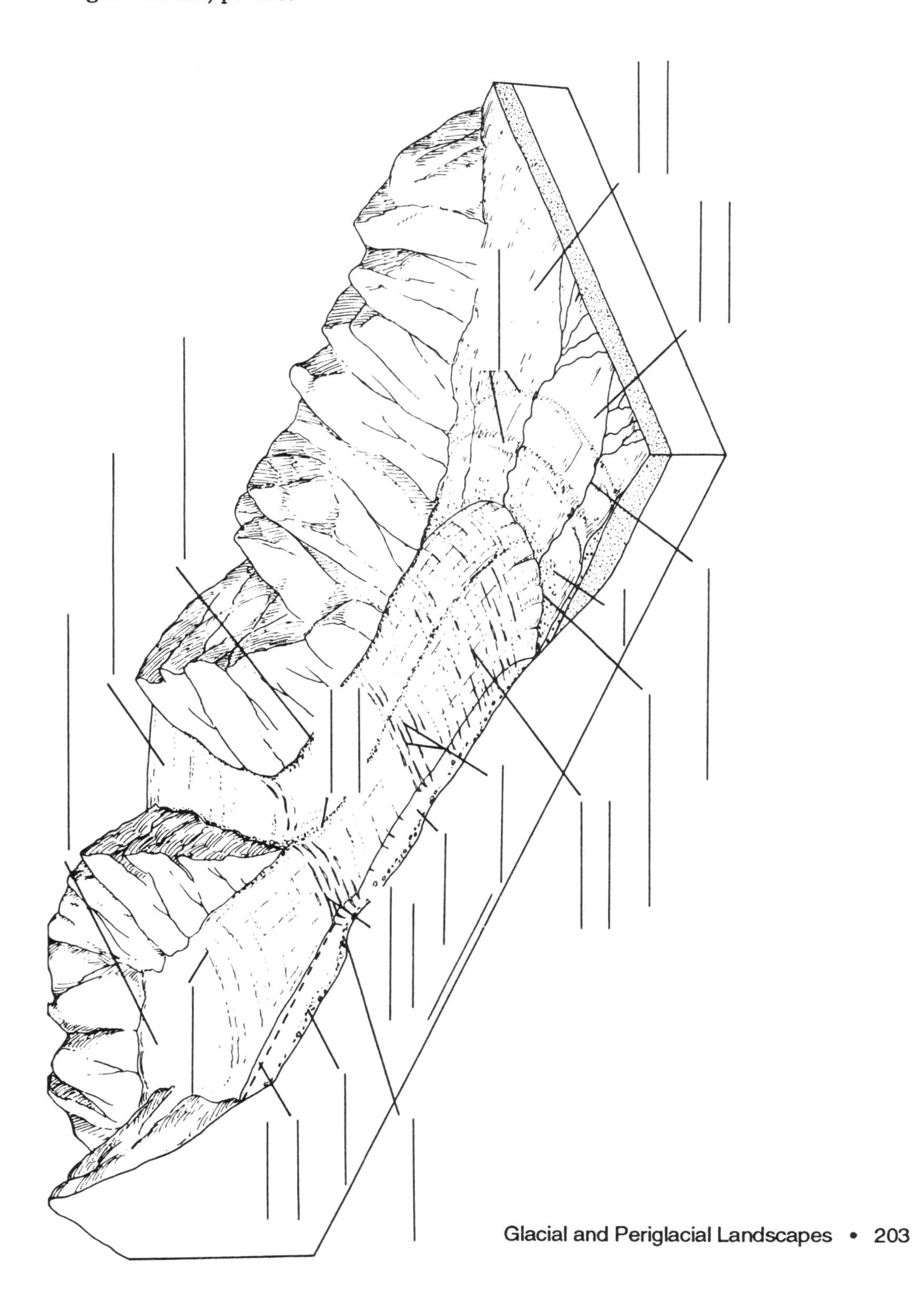

7. What dynamic changes are occurring in the South Cascade Glacier (News Report #1, p. 428)?

____

____

____.

8. Describe what is meant by "glacial surge" and give at least two examples of this type of glacial movement.

____

____

____.

(a) What is shown in the *Landsat* image in Figure 14-5, p. 430?________________

____.

9. Using Figure 14-7a, b, and c (p. 518), characterize as completely as possible with written descriptions each of the stages of a landscape subjected to fluvial (a) and glacial processes and actions (b and c).

(a)____

____

____.

(b)____

____

____.

(c)____

____

____.

10. List the features common to alpine (valley) glaciation that are absent in continental glaciation (see Table 17-1, p. 522).__________________________________

____

____

____.

11. "Topographic Map 4 (T.4): Mount Rainier National Park" is included in the back of your text after the Index. The area shown on the map segment is a portion of the tallest volcano in the Cascade Range that stretches from the Canadian border to northern California. Using this map and the discussion in the text, answer the following. The legend for topographic map symbols is in Appendix B, p. B.2.

(a) What is the contour interval for this topographic map?______________________________.

(b) Name several glaciers on which you identify medial and lateral moraines____________

____

____.

(c) Tributary streams are left stranded high above the glaciated valley floor, following the removal of previous slopes and stream courses by the glacier. These *hanging valleys* are the sites of spectacular waterfalls as streams plunge down the steep cliffs. Locate several waterfalls on the map segment, name the stream and the falls:

____

____.

(d) The summit of Mount Rainier is at 14,410 ft (4392 m). What is the lowest elevation that you can find on this map? ____.

Where is this location?____

____.

12. Using your text and Figure 14-7 b and c, (p. 431) complete the labels on the following illustration.

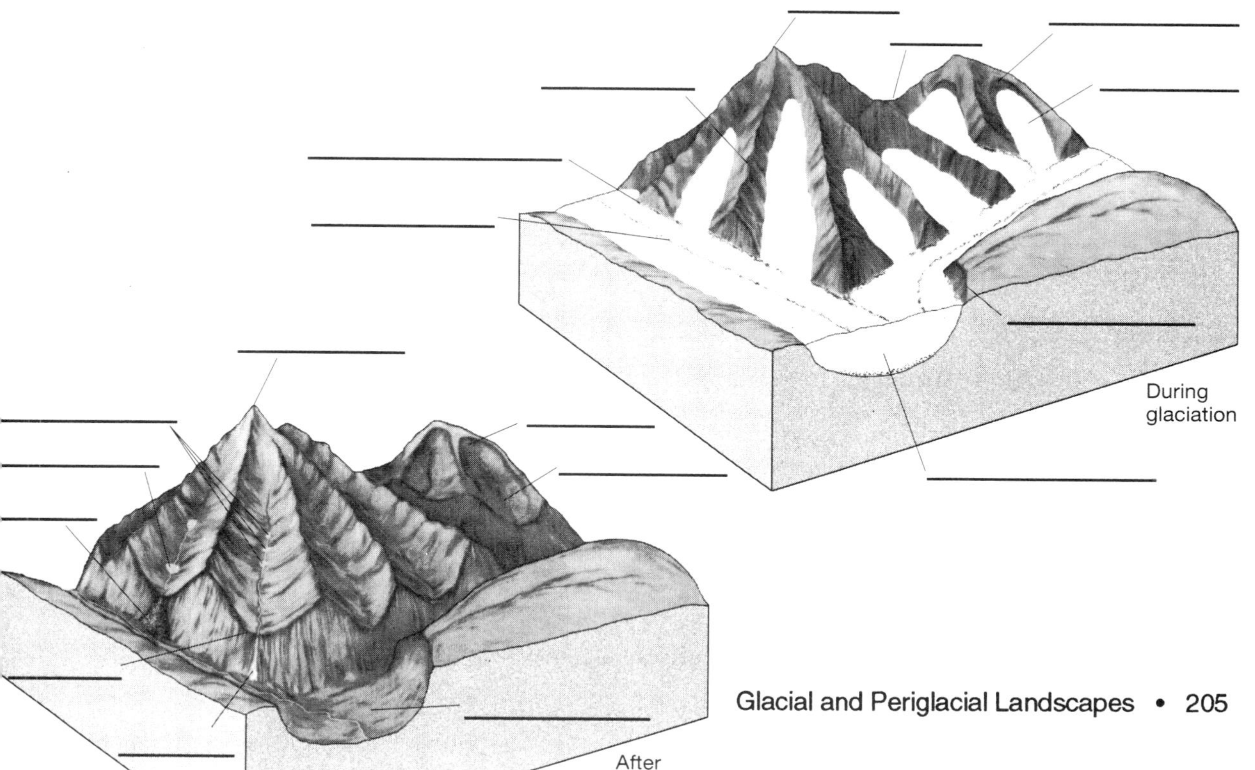

13. Compare and contrast the following two landforms produced by continental glaciation.

(a) roche moutonnée:________________________________

____

____

Does this description fit the photograph in Figure 14-12, p. 436? Explain.________________

(b) drumlin:________________________________

____

____

Can you identify these features on the topographic map in Figure 14-13, p. 436?

____

14. "Topographic Map 5 (T.5): Jackson, Michigan" is included in the back of your text after the Index. The area shown on the map segment is a region of the United States that was blanketed several times by continental glaciers. Using this map and the discussion in the text, answer the following. The legend for topographic map symbols is in Appendix B.

(a) What is the contour interval on this topographic map?________________

(b) Using the text and Figure 14-11, p. 435, can you identify what type of landform is labeled "Blue Ridge" ?________________.

(c) The features called Mud Lake, Skiff Lake, and Crystal Lake occupy indentations called ________________and result from isolated blocks of ice left behind as the glaciers retreated.

15. Define each and distinguish between the following terms:

(a) permafrost:________________________________

____

(b) ground ice:________________________________

____

(c) patterned ground:________________________________

____

16. List two examples of ground ice formations or landforms.

(a) ________________________________

(b) ________________________________

17. Describe what is depicted in Figure 14-17? Why is the cabin failing? The construction of "Utilidors" shown in Figure 14-18 is necessary because of what conditions? Relate these in your answer.

____

____

____

____.

18. Relative to the text discussion of glacial and interglacial stages and Figure 14-19, p. 441, describe the following.

(a) Illinoian (number of stages ?):______________________________

____.

(b) Wisconsinan (number of stages ?):______________________________

____.

(c) Sangamon:______________________________

____.

(d) Stage 23:______________________________

____.

19. Describe the physical criteria used to delimit the polar regions.

(a) Arctic:______________________________

____.

(b) Antarctic:______________________________

____.

20. In terms of paleoclimatology and past climatic changes over the long span of geologic history, what three physical factors are mentioned as probable causes?

(a)______________________________

(b)______________________________

(c)______________________________

Sample Self-test

(Answers appear at the end of the study guide.)

1. A general term for a mass of perennial ice, resting on land or floating shelflike in the sea adjacent to land, is a/an
   a) snowline
   b) iceberg
   c) glacier
   d) ice field

2. Alpine glaciers include all of the following except a/an
   a) ice cap
   b) mountain glacier
   c) cirque glacier
   d) valley glacier

3. Relative to glacial mass balance, which of the following is <u>incorrect</u>?
   a) a positive net balance or negative net balance occurs during a cold period and warm period respectively
   b) glacial mass is reduced by evaporation, sublimation, and deflation
   c) glacial mass is reduced by a combination of processes called ablation
   d) the zone in the glacier where accumulation gains and losses begin is the equilibrium line
   e) worldwide, most glaciers are showing marked increases in mass at present

4. Glacial ice is
   a) essentially the same as snow
   b) formed after a long process that may take 1000 years in Antarctica
   c) also known as firn
   d) generally less dense than snow and firn

5. Glacial erosion specifically involves
   a) ablation
   b) deflation
   c) abrasion and plucking
   d) eskers and kames

6. Periglacial processes
   a) affect approximately 20% of Earth's land surface
   b) occur at high latitudes and high altitudes
   c) involve permafrost, frost action, and ground ice
   d) all of these are correct

7. Which of the following is <u>incorrect</u> relative to the Pleistocene ice age epoch?
   a) it began 1.65 million years ago
   b) it produced ice sheets and glaciers that covered 30 percent of Earth's land area
   c) at least 18 expansions of ice occurred over Europe and North America
   d) it represents a single continuous cold spell

8. Geomorphic landforms produced by valley glaciers include which of the following?
   a) drumlins
   b) eskers
   c) ice caps and sheets
   d) horns and cols

9. In terms of paleoclimatology, which of the following was not mentioned as a probable cause of past ice ages?
   a) the weight of the ice
   b) galactic and Earth-Sun relationships
   c) geophysical factors
   d) geographical-geological factors

10. Milankovitch proposed astronomical factors as a basic cause of ice ages.
   a) true
   b) false

11. The Great Salt Lake, a remnant of a paleolake (pluvial lake), reached record historic levels during the 1980s.
   a) true
   b) false

12. The definition of the Antarctic region is simply the extent of the Antarctic continental landmass.
   a) true
   b) false

13. Greenland actually contains more ice than does Antarctica at the present time.
   a) true
   b) false

14. A valley glacier is commonly associated with unconfined bodies of ice.
   a) true
   b) false

15. The Vatnajökull of Iceland is presented as an example of an ice cap.
   a) true
   b) false

16. Internally, a glacier can continue to move forward even though its lower terminus is in retreat.
   a) true
   b) false

17. A **V**-shaped valley is characteristic of glaciated valleys, whereas a **U**-shape is more characteristic of a stream-eroded valley.
   a) true
   b) false

18. An ice wedge is a form of ground ice in periglacial landscapes.

a) true
b) false

19. Glacial till is unsorted, whereas glacial outwash is sorted and stratified as one would expect stream-deposited materials to be.

a) true
b) false

20. A glacier may have several end moraines but only one terminal moraine.

a) true
b) false

21. The cirques where the valley glacier originated form cirque walls that wear away; when this happens, an ________________, or a sharp ridge that divides two cirque basins forms. When two eroding cirques reduce this to form a pass or saddlelike depression, the term________________ is used. A pyramidal peak called a________________ results when several cirque glaciers gouge an individual mountain summit from all sides. A small mountain lake, especially one that collects in a cirque basin behind risers of rock material, is called a ________________. Small, circular, stair-stepped lakes in a series are called ________________ because they look like a string of rosary (religious) beads forming in individual rock basins aligned down the course of a glaciated valley.

22. The scientist credited with the theory that Europe was once covered by blankets of ice is

________________________________(include lifetime dates).

PART FOUR: Biogeography

Overview–Part Four

Earth is the home of the only known biosphere in the Solar System: a unique, complex, and interactive system of abiotic and biotic components working together to sustain a tremendous diversity of life. Thus we begin Part Four of ELEMENTAL GEOSYSTEMS and an examination of the geography of the biosphere—soils, ecosystems, and terrestrial biomes. Recall the description of the biosphere given in Chapter 1:

> **The intricate web that connects all organisms with their physical environment is the biosphere. Sometimes referred to as the ecosphere, the biosphere is the area in which physical and chemical factors form the context of life. The biosphere exists in an area of overlap among the spheres. Life processes have also powerfully shaped the other spheres through various interactive processes. The biosphere has evolved, reorganized itself at times, faced extinction, gained new vitality, and managed to flourish overall. Earth's biosphere is the only known one in the solar system; thus, life as we know it is unique to Earth.**

Part Four is a synthesis of many of the elements covered throughout the text.

Name:____________________________ Class Section:________________

Date:__________________ Score/Grade:____________

The Geography of Soils

15

Chapter Overview

Earth's landscape is generally covered with soil. Soil is a dynamic natural body and comprises fine materials, in which plants grow, and which is composed of both mineral and organic matter. Because of their diverse nature,soils are a complex subject and pose a challenge for spatial analysis. This chapter presents an overview of the modern system of soil classification used in the United States, with mention of the Canadian system.

Please find Table 15-1, pp. 464-65, and utilize it to simplify this geographic study of soils.

The table summarizes the eleven soil orders of the Soil Taxonomy. The table includes general location and climate association, areal coverage estimate, and basic description. Also, for most soil orders, you will find small locator maps included, along with a picture of that soil's profile. These locator maps are derived from the worldwide distribution map presented in Figure 15-8, pp. 462-63.

Learning Objectives

The following learning objectives help guide your reading and comprehension efforts. The operative word is in *italics*. Read and work with these carefully and note that exercise #1 asks you about five of these objectives. After reading the chapter and using this workbook, you should be able to:

1. *Discern* the importance of soil as a dynamic natural body.
2. *Define* the concept of soil fertility and *relate* this to productivity.
3. *Describe* a pedon and a polypedon.
4. *Portray* soil horizons and *define* each component in a typical soil profile.
5. *Define* humus and *relate* this to the soil profile.
6. *Describe* the soil erosion process of eluviation and the soil depositional process of illuviation.
7. *Define* solum.
9. *Relate* soil color to soil properties.
10. *Describe* loam and *relate* it to soil texture.
11. *Utilize* soil structure, consistence, and porosity to analyze soil properties.
12. *Relate* soil moisture characteristics to the water balance approach introduced in Chapter 6.
13. *Explain* basic soil chemistry including cation-exchange capacity and soil pH.
14. *Describe* three basic soil-formation factors: dynamic, passive, and human.
15. *Evaluate* human management and mismanagement of soil and *explain* the impact of human intervention in producing soil erosion.
16. *Relate* these erosion and soil degradation concerns to the global distribution portrayed on the world map in Figure 15-7.
17. *Outline* a brief history of soil classification in the United States and *identify* the elements that produced present classification schemes.
18. *Differentiate* and *identify* essential epipedons and subsurface diagnostic horizons.
19. *Define* the factors in the Soil Taxonomy classification system.
20. *List* the eleven soil orders of the Soil Taxonomy and *relate* their occurrence worldwide where applicable.
21. *Explain* pedogenic regimes and *describe* three such processes as they relate to specific soil orders: laterization, calcification, and salinization.
22. *Describe* selenium concentrations in western soils and *explain* the implications of these occurrences.

Outline Headings and Glossary Review

These are the first- second-, and third-order headings that divide this chapter. The key terms and concepts that appear **boldface** in the text are listed under their appropriate heading in bold italics; these highlighted terms appear in the text glossary. A check-off box is placed next to each key term so you can mark your progress through the chapter as you define these in your reading notes or prepare note cards.

- ❑ ***soil***
- ❑ ***soil fertility***

Soil Characteristics

Soil Profiles

- ❑ ***pedon***
- ❑ ***polypedon***

Soil Horizons

- ❑ ***soil horizon***
- ❑ ***humus***
- ❑ ***eluviation***
- ❑ ***illuviation***

- ❑ ***solum***

Soil Properties
- Soil Color
- Soil Texture
  - ❑ ***loam***
- Soil Structure
- Soil Consistence
- Soil Porosity
- Soil Moisture

Soil Chemistry
- Retaining Ions for Plants
  - ❑ ***soil colloids***
  - ❑ ***adsorption***
  - ❑ ***cation-exchange capacity (CEC)***
- Soil Acidity and Alkalinity

Soil Formation Factors and Management
- Natural Factors
- The Human Factor

Soil Classification

A Brief History of Soil Classification
- ❑ ***Soil Taxonomy***
- U. S. Soil Taxonomy

Diagnostic Soil Horizons
- ❑ ***epipedon***
- ❑ ***subsurface diagnostic horizon***

The Eleven Soil Orders of Soil Taxonomy

Oxisols
- ❑ ***Oxisols***
- ❑ ***laterization***
- ❑ ***plinthite***

Aridisols
- ❑ ***Aridisols***
- ❑ ***salinization***

Mollisols
- ❑ ***Mollisols***
- ❑ ***calcification***
- ❑ ***caliche***

Alfisols
- ❑ ***Alfisols***

Ultisols
- ❑ ***Ultisols***

Spodosols
- ❑ ***Spodosols***
- ❑ ***podzolization***

Entisols
- ❑ ***Entisols***

Inceptisols
- ❑ ***Inceptisols***

Andisols
- ❑ ***Andisols***

Vertisols
- ❑ ***Vertisols***

Histosols
- ❑ ***Histosols***

SUMMARY

Learning Activities and Critical Thinking

1. Select any five learning objectives from the list presented at the beginning of this chapter. Place the number selected in the space provided (no need to rewrite each objective). Using the following questions as guidelines only, briefly discuss your treatment of the objective.

- What did you know about the objective before you began?
- What was your plan to complete the objective?
- Which information source did you use in your learning (text, or other)?
- Were you able to complete the action stated in the objective? What did you learn?
- Are there any aspects of the objective about which you want to know more?

a)_____:____

____

____

b)_____ :

c)_____ :

d)_____ :

e)_____ :

2. Describe the chapter-opening photograph for Chapter 15:

3. Define the horizons of a soil profile.

(a) O:____

(b) A:____

(c) E:____

(d) B:____

(e) C:____

(f) R:____

4. Label and describe each specific horizon in the following illustration taken from Figure 15-2 (p. 455).

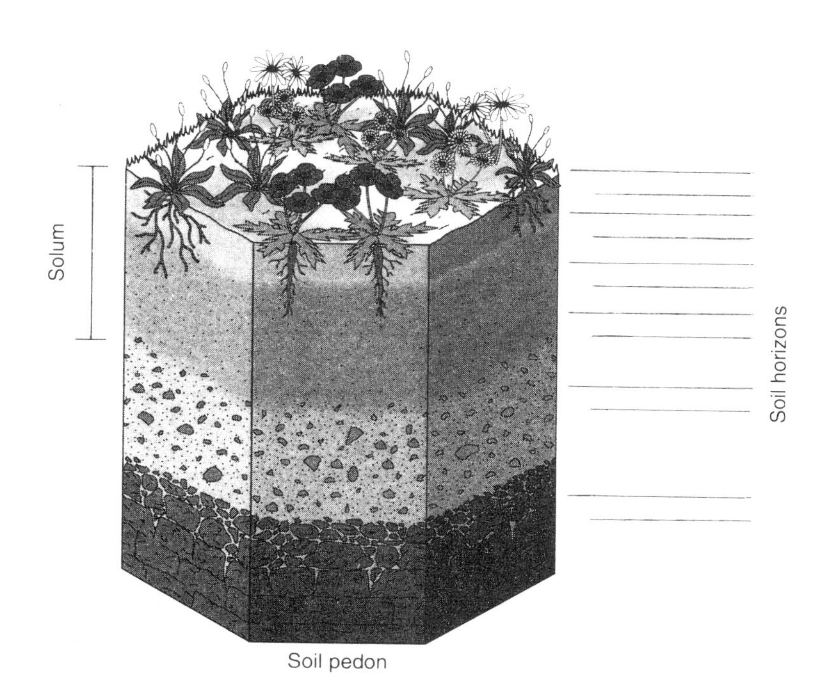

5. As completely as possible, define **humus:**____

6. What is meant by the key concept "**soil fertility**"? Explain.______________________________

7. Using the soil texture diagram in Figure 15-3 (p. 457), identify the textural analysis of a Miami silt loam soil described in the text. Locate each of the A, B, and C horizons on the illustration and draw a line from the three related axes of the triangle to the properties of the three horizons so that these textural analyses lines intersect. The sampling points are marked 1, 2, and 3 on the illustration.

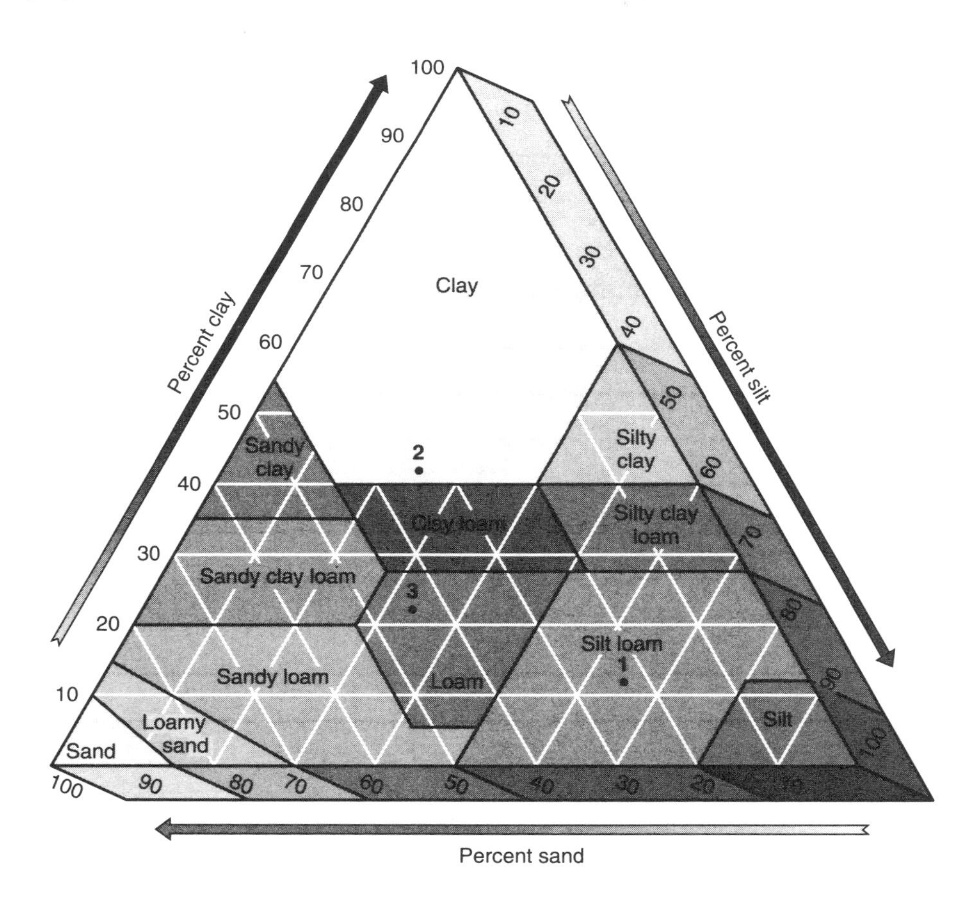

8. Define the following five terms related to soil moisture using the information presented in both Chapter 15 and in Chapter 6, pp. 188-93.

(a) soil moisture utilization:____

____

____.

(b) soil moisture recharge:____

____

____.

(c) field capacity:____

____

____.

(d) wilting point:____

____

____.

(e) capillary (available) water:______________________________________

____

____.

9. Relative to soil chemistry, define and characterize each of the following.

(a) pH:____

____.

(b) alkalinity:____

____.

(c) acidity:____

____.

10. Explain soil erosion. Is it a problem at this time in the United States? Canada? Worldwide? If so, why is this a potentially severe situation? What is the magnitude of this issue as characterized in the text (p. 460), News Report #1 (p. 466), and map in Figure 15-7 (p. 461)?

____

____

____

____

____.

11. Relative to News Report #3, "The Dust Bowl," p. 382 in Chapter 12, relate the circumstances of this historic soil disaster. Do you discern techniques or methods that might prevent such a catastrophe in the future, or was the dust bowl purely a natural happening over which we have no control?

____

____

____

____

____.

12. Using Table 15-1 (pp. 464-65) , world map in Figure 15-8, and text pages 466-78, give basic geographic description of representative locations for each soil order. You might want to complete this section while using the glossary checklist at the beginning of the chapter..

Oxisols:______________________________________

____

____.

Aridisols:_____________________________________

____

____.

Mollisols:_____________________________________

____

____.

Alfisols:______________________________________

____

____.

Ultisols:______________________________________

____

____.

Spodosols:____________________________________

____

____.

Entisols:____

____

____.

Inceptisols:____

____

____.

Andisols:____

____

____.

Vertisols:____

____

____.

Histosols:____

____

____.

13. Explain and analyze the "Death of the Kesterson National Wildlife Refuge" (News Report #2, p. 470).

____

____

____

____

____

____

____.

Sample Self-test

(Answers appear at the end of the study guide.)

1. Soil is
   a) a physical, geological product
   b) ground-up rock
   c) a dynamic natural body of mineral and organic matter
   d) composed of mineral particles
   e) not a result of weathering processes

2. The basic sampling unit in soil surveys is called a/an
   a) polypedon
   b) pedon
   c) A horizon
   d) horizon

3. The R horizon refers to
   a) weathered or unconsolidated bedrock
   b) soils redeveloped through reclamation
   c) a zone of eluviation in the soil horizon
   d) the basic mapping unit of soils in an area

4. Soil Taxonomy refers to
   a) a method of soil classification based on actual soil properties
   b) a system that has been in use since 1938
   c) a system of classification that we owe to Dukuchaev and the Russians
   d) the Marbut classification system

5. Which of the following lists is in the proper hierarchy of the Soil Taxonomy system (least to most occurrences)?
   a) orders, suborders, soil series, small groups
   b) soil orders, series, great horizons, small horizons
   c) soil orders, suborders, great groups, subgroups, families, series
   d) pedon, polypedon, horizons, profiles

6. A soil order characteristic of the Amazon Basin is a/an
   a) Alfisol
   b) Aridisol
   c) Spodosol
   d) Oxisol

7. Salinization is a specific process roughly associated with
   a) Ultisols
   b) Histosol
   c) Aridisols
   d) Mollisols, soils of central Iowa,

8. When limestone is added to Spodosols,
   a) soil pH is lowered to benefit fertility
   b) it firms up a weak soil structure
   c) it adds important colloids to the soil profile
   d) it raises pH, raising crop yields per acre

9. The undeveloped soils of Zabriskie Point, Death Valley, are characteristic Entisols.
   a) true
   b) false

10. Inceptisols are typical of arctic tundra regions.
   a) true
   b) false

11. Bogs of sphagnum peat are not related to any soil order.
   a) true
   b) false

12. The Kesterson Wildlife Refuge was destroyed by salinization and concentrations of selenium.
   a) true
   b) false

13. Each layer exposed in a pedon is called a soil horizon.
   a) true
   b) false

14. The B horizon normally experiences the greatest concentration of organic matter.
   a) true
   b) false

15. Individual mineral particles are called peds.
   a) true
   b) false

16. Individual soil aggregates are called soil separates.
   a) true
   b) false

17. About 13,000 soil types (soil series) have been recognized in the United States alone.
   a) true
   b) false

18. Soil erosion may remove a few centimeters of soil a year—an amount that may have taken 500 years to form.
   a) true
   b) false

19. Andisols represent a relatively new classification (1990) for soils derived from volcanic sources.
   a) true
   b) false

20. Histosols represent organic soils such as those you find in a sphagnum peat bog.
   a) true
   b) false

Name:______________________________ Class Section:____________________

Date:____________________ Score/Grade:______________

Ecosystems and Biomes

16

Chapter Overview

The interaction of the atmosphere, hydrosphere, and lithosphere produces conditions within which the biosphere exists. Chapter 16 continues the process of synthesizing all these "spheres" into a complete spatial picture of Earth. In this complex age, the spatial tools of the geographic approach are uniquely suited to unravel the web of human impact on Earth's systems. Many career opportunities in planning, GIS analysis, environmental impact assessment, and location analysis are available to those with a degree in geography.

The biosphere extends from the floor of the ocean to a height of about 8 km (5 mi) into the atmosphere. The biosphere is composed of myriad ecosystems, from simple to complex, each operating within general spatial boundaries. Ecology is the study of the relationships between organisms and their environment and among the various ecosystems in the biosphere. Biogeography, essentially a spatial ecology, is the study of the distribution of plants and animals and the diverse spatial patterns they create across Earth.

Chapter 16 synthesizes many of the elements of ELEMENTAL GEOSYSTEMS, bringing them together to create a regional portrait of the biosphere. To facilitate this process, Table 16-2, p. 508-09, is presented to portray aspects of the atmosphere, hydrosphere, and lithosphere that merge to produce the major terrestrial ecosystems. The table contains columns of vegetation characteristics, soil classes, Köppen climate designation, annual precipitation range, temperature patterns, and water balance characteristics. Biomes possess cures and clues to human disease and mechanisms to recycle the excessive levels of carbon dioxide now entering the atmosphere.

The biosphere is quite resilient and adaptable, whereas many of the specific biomes and communities are greatly threatened by further destructive impacts by society. The irony is that some plants and animals contain cures and clues to human disease, are potential new food sources, and, for plants, operate as mechanisms that recycle excessive carbon dioxide now entering the atmosphere.

We are the species with developed technology that allows us to lift off the surface of our home planet despite the protests of gravity, to view Earth from afar. At a moment when these accomplishments fill us with pride, we also find ourselves overwhelmed with the immensity of the home planet and the smallness of our everyday reality, as revealed to us by photographs of Earth—such as the one on the back cover of this textbook. A proper understanding of this chapter on global ecosystems is best fueled by an inner picture of our Earth stimulated by the one taken from space, for it is only from that perspective that the importance of even the smallest community, or the most finite ecosystem, will arise in proper significance. And, from such awareness even the loss of a single species will be a headline and a topic of concern to us all.

Learning Objectives

The following learning objectives help guide your reading and comprehension efforts. The operative word is in *italics*. Read and work with these carefully and note that exercise #1 asks you about five of these objectives. After reading the chapter and using this workbook, you should be able to:

1. *Define* biosphere and *relate* the definition to Earth's myriad ecosystems.
2. *Compare* and *contrast* ecology and biogeography and *discern* the spatial nature of the geographic approach.
3. *Interpret* the quotes from Gilbert White on p. 482 and in the caption to Figure 16-1(p. 483) from Amory Lovins and *relate* these to the complexity of the biosphere.
4. *Define* and *differentiate* ecosystems and communities.
5. *Define* habitat and niche and *assess* the possible relationship of these concepts to humans.
6. *Explain* photosynthesis and respiration and *derive* net photosynthesis.
7. *Analyze* the world pattern of net primary productivity and *discern* regions of high productivity and low productivity.
8. *List* abiotic ecosystem components and *relate* these to ecosystem operations.
9. *Outline* the abiotic climate control of ecosystem types and *explain* the generalized relationship among rainfall, temperature, and vegetation.
10. *Describe* the life zone concept and *portray* a typical plant zonation with altitude.
11. *Describe* several biogeochemical cycles in nature: oxygen and carbon dioxide, and nitrogen.
12. *Explain* the limiting factor concept and *illustrate* this with several examples.
13. *List* biotic ecosystem components and *relate* these to ecosystem operations.
14. *Differentiate* between a food chain and food web and *illustrate* this using krill in Antarctic waters.
15. *Relate* the biomass pyramid concept to North American dietary habits and food chain efficiency.
16. *Explain* the trophic relationships in a grassland and in a temperate forest during typical summer conditions.
17. *Explain* ecosystem stability and *relate* this to possible conditions of instability.
18. *Describe* the impact of climate change on the distribution of plant species and *illustrate* this potential with a specific example.
19. *Define* succession and *outline* the stages of general ecological succession in both terrestrial and aquatic ecosystems.
20. *Discern* the importance of regions where shared abiotic and biotic traits dominate.
21. *Review* the large marine ecosystem (LMEs) concept and *list* several representative examples including the largest of the protected areas in North America (News Report #3, p. 503).
22. *Define* six formation classes and *explain* their relationship to plant communities.
23. *Define* life-form designations as a system used for the structural classification of plants.
24. *Define* the terrestrial ecosystem concept and *identify* ten major biomes.
25. *Relate* each major biome to vegetation characteristics, soil classes, Köppen climate designation, annual precipitation range, temperature patterns, and water balance characteristics (Table 16-2).
26. *Locate* each of the major biomes on the world map in Figure 16-22, pp. 506-07, and *relate* each section in the chapter to the map as you read about each biome.
27. *Compare* and *correlate* the terrestrial biome map to the climate classification map inside the front cover of this text.

28. *Analyze* the ongoing deforestation of the equatorial and tropical rain forests and *describe* the implications of these losses.
29. *Portray* the biosphere reserve approach to species preservation and *relate* the potential loss of species we face.
30. *Discern* the geographic aspects of taiga and boreal forest.
31. *Review* the different terms used worldwide to describe Mediterranean shrublands.
32. *Describe* how much of the original grassland remains in North America.
33. *Identify* a few of the unique adaptations of life forms in the desert biome.
34. *Differentiate* between warm and cold desert and semidesert regions.

Outline Headings and Glossary Review

These are the first- second-, and third-order headings that divide this chapter. The key terms and concepts that appear **boldface** in the text are listed under their appropriate heading in bold italics; these highlighted terms appear in the text glossary. A check-off box is placed next to each key term so you can mark your progress through the chapter as you define these in your reading notes or prepare note cards.

❑ ***ecosystem***
❑ ***ecology***
❑ ***biogeography***

Ecosystem Components and Cycles

Communities
❑ ***community***
❑ ***habitat***
❑ ***niche***

Plants: The Essential Biotic Component
Beginnings
Leaf Activity
❑ ***stomata***

Photosynthesis and Respiration
❑ ***photosynthesis***
❑ ***chlorophyll***
❑ ***respiration***

Net Primary Productivity
❑ ***net primary productivity***
❑ ***biomass***

Abiotic Ecosystem Components
Light, Temperature, Water, and Climate
❑ ***photoperiod***
❑ ***life zone***

Gaseous and Sedimentary Cycles
❑ ***biogeochemical cycles***

Limiting Factors
❑ ***limiting factor***

Biotic Ecosystem Operations
Producers, Consumers, and Decomposers
❑ ***producers***
❑ ***consumers***
❑ ***food chain***
❑ ***food web***
❑ ***herbivore***
❑ ***carnivore***
❑ ***omnivore***
❑ ***decomposers***

Ecological Relationships
Concentration in Food Chains

Stability and Succession

Ecosystem Stability and Diversity
❑ ***diversity***

Agricultural Ecosystems
Climate Change
Ecological Succession
❑ ***ecological succession***

Terrestrial Succession
Fire Ecology
Aquatic Succession
❑ ***eutrophication***

Earth's Ecosystems

❑ ***aquatic ecosystems***

Terrestrial Ecosystems
❑ ***terrestrial ecosystem***
❑ ***biome***
❑ ***formation classes***
❑ ***ecotone***

Earth's Major Terrestrial Biomes

Equatorial and Tropical Rain Forest
❑ ***equatorial and tropical rain forest***

Deforestation of the Tropics
Tropical Seasonal Forest and Scrub
❑ ***tropical seasonal forest and scrub***

Tropical Savanna

❑ ***tropical savanna***

Midlatitude Broadleaf and Mixed Forest

❑ ***midlatitude broadleaf and mixed forest***

Northern Needleleaf Forest and Montane Forest

❑ ***northern needleleaf forest***

❑ ***boreal***

❑ ***taiga***

Temperate Rain Forest

❑ ***temperate rain forest***

Mediterranean Shrubland

❑ ***Mediterranean shrubland***

❑ ***chaparral***

Midlatitude Grasslands

❑ ***midlatitude grasslands***

Desert Biomes

❑ ***desert biomes***

Desertification

Warm Desert and Semidesert

Cold Desert and Semidesert

Arctic and Alpine Tundra

❑ ***arctic tundra***

SUMMARY

Learning Activities and Critical Thinking

1. Select any five learning objectives from the list presented at the beginning of this chapter. Place the number selected in the space provided (no need to rewrite each objective). Using the following questions as guidelines only, briefly discuss your treatmen of the objective.

- What did you know about the objective before you began?
- What was your plan to complete the objective?
- Which information source did you use in your learning (text, or other)?
- Were you able to complete the action stated in the objective? What did you learn?
- Are there any aspects of the objective about which you want to know more?

a)____:______________________________

____

____

____

____

____

b)____:______________________________

____

____

____

____

____

c)____ :____

____

____

____

____

____.

d)____ :____

____

____

____

____

____.

e)____ :____

____

____

____

____

____.

2. Describe lichen as an example of the principle of *symbiosis*. Is there any aspect of this relationship that might be analogous to the biosphere and Earth-human relationships?

____

____

____

____.

3. In the space provided, record the formulas for *photosynthesis* and *respiration* (p. 485). Identify each component by symbol and with its name placed below on the next line.

(a) photosynthesis:______________________ = ______________________.

____

(b) respiration:______________________ = ______________________.

____

4. Complete the labeling for this illustration of abiotic and biotic system components as derived from Figure 16-2 (p. 483). You may want to augment the illustration with color shading.

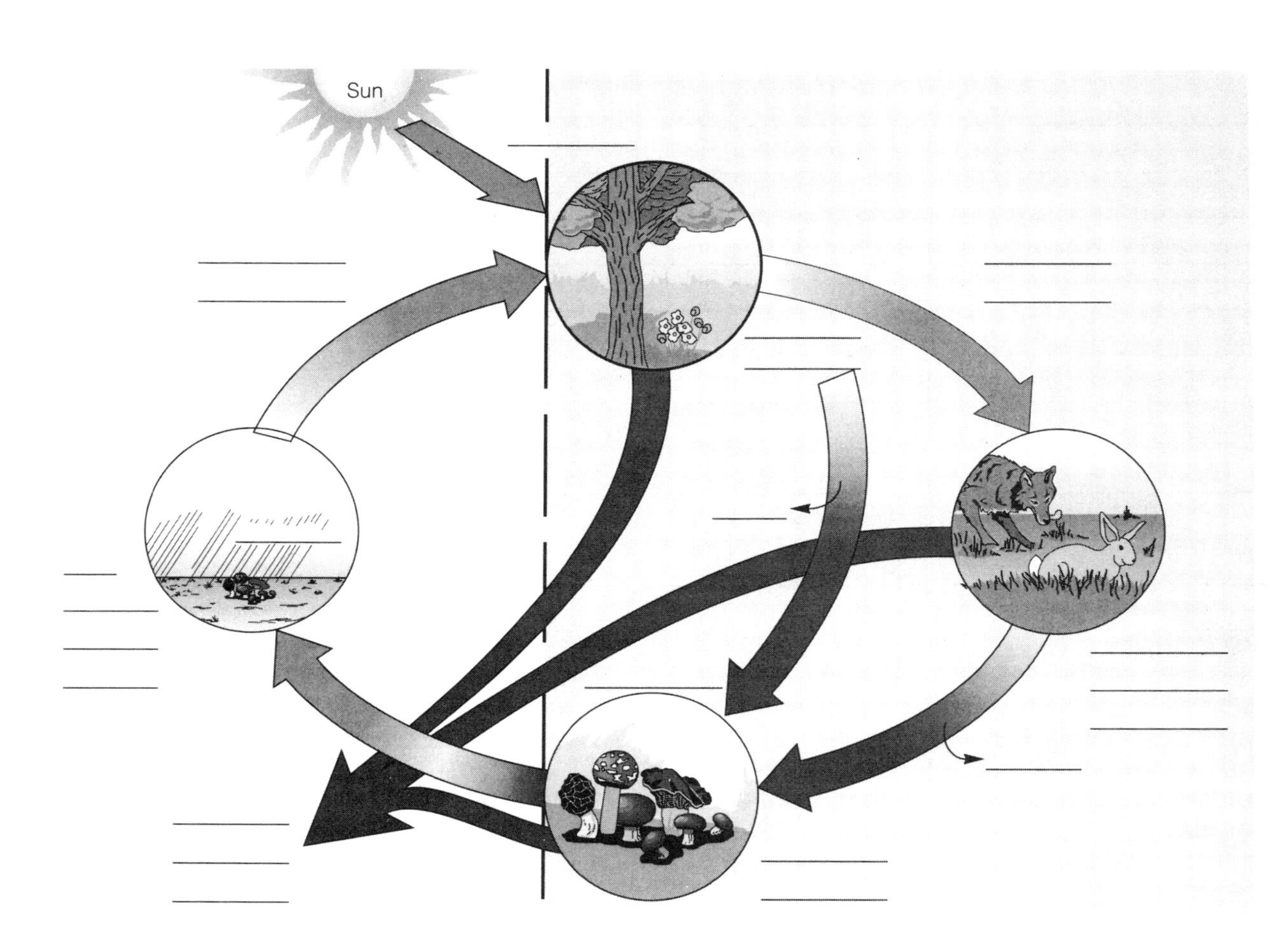

5. Describe in general terms the distribution of worldwide net primary productivity shown in Figure 16-5, p. 487, and discussed in the text (p. 486). Relate this distribution to any climatic factors that may assist in explaining the distribution.

____

____

____

____

6. List and compare the net primary production and plant biomass on Earth for the following ecosystems (Table 16-1, p. 487):

Ecosystem	Area ($10^6 km^2$)	Net Primary Productivity per Unit Area ($g/m^2/yr$) Normal Range Mean	World Net Primary Production ($10^9 t/yr$)
Tropical rain forest			
Temperate deciduous			
Boreal forest			
Savanna			
Tundra			
Desert and semidesert			
Cultivated land			
Algal beds and reefs			
Estuaries			

7. As abiotic ecosystem components, briefly describe the importance of each of these physical factors.

(a) light:____

____.

(b) temperature:__________________________________

____.

(c) water:____

____.

(d) climate:______________________________________

____.

8. What does the illustration in Figure 16-6, p. 488, tell you about the relationship among these physical components in determining ecosystem types?

____

____

____

____

____

____

____.

9. In the space provided, make a simple sketch of the relationships among producers, consumers, and decomposers in a typical food chain. Identify each trophic level and activity in your sketch with appropriate labels. Begin by placing primary producers at the lower margin of the box. (Use the discussion in pp. 493-96 and Figures 16-12, 16-13, and 16-14 for basic information.)

10. How do you explain the great difference in areal distribution between mallards and snail kites as depicted in Figure 16-11b, p. 494?

____

____

____

____.

11. Label and detail the following illustration of carbon and oxygen cycles derived from Figure 16-9, p. 491.

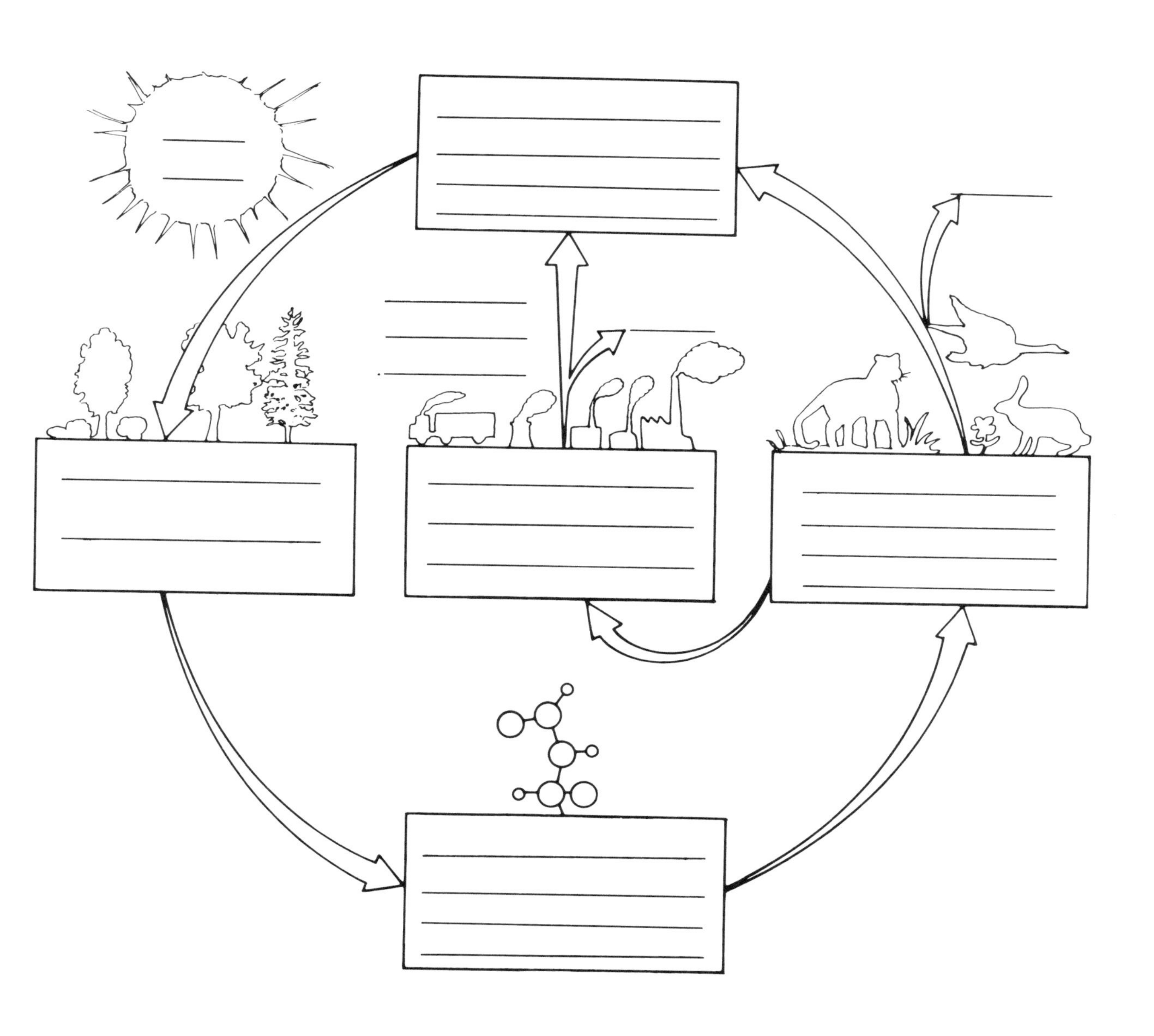

12. What specifically is meant by "Humans Dump Carbon into the Atmosphere" (News Report #1, p. 493)? Tie-in your answer to Table 7-1, p. 240. Identify the consequences of this situation.

____

____

____

____

____.

13. In basic terms, describe the process of ecological succession in a community of plants and animals as suggested in Figure 16-17, 16-18, pp. 499-500, and the text.

____

____

____

____.

14. Explain the principle of fire ecology and succession. Relate what happened at Yellowstone in 1988 (News Report #2, p. 501.______________________________

____

____

____.

(a) How have modern fire prevention skills actually worsened the threat of wildfires?

____

____

____.

15. Describe the concept and reality of "Large Marine Ecosystems" (News Report #3, p. 503). ____

____

____

____

____.

16. Explain the significance of biogeographical realms and their relationship to establishing regional patterns of Earth's biomes.

____

____

____.

17. As a review, *reread* and *recheck* your notes for the definitions of the following terms on pp. 482-84: *ecosystem*, *ecology*, *biogeography*, *community*, *habitat*, and *niche*.

18. Given the description of the equatorial and tropical rain forest (pp. 505-511), examine the chapter-opening photograph taken in Puerto Rico. Can you identify any of the features described by the text in the photograph? Explain.

____

____

____.

19. Present a brief overview of the condition and status of deforestation in the tropics (p. 511-12, Figure 16-25, and News Report #4, p. 516).

____

____

____

____.

20. From Table 16-2, pp. 508-09, and the map in Figure 16-22, pp. 506-07, determine the biome that best characterizes each of the following descriptions and record its name in the space provided:

An annual precipitation range of 25-75 cm:________________________________.

POTET greater than 1/2 PRECIP:________________________________.

Southern and eastern evergreen pines:________________________________.

Mollisols and Aridisols:________________________________.

Temperate with a cold season:________________________________.

Characteristic of central Australia:________________________________.

Selva:____

____.

Sclerophyllous shrub:________________________________.

Short summer, cold winter:____________________.

Characteristic of the majority of central Canada:____________________.

Transitional between rain forest and tropical steppes:____________________.

Tallest trees on Earth:____________________.

Sedges, mosses, and lichens:____________________.

Characteristic of Zambia (south central Africa):____________________.

Four biome types that occur in Chile:____________________

____________________.

Less than 40 rainy days in summer:____________________.

Characteristic of Tennessee:____________________.

Precipitation of 150-500 cm/year, outside the tropics:____________________.

Always warm, water surpluses all year:____________________.

Characteristic of central Greenland:____________________.

Characteristic of Iran (northeast of the Persian Gulf):____________________.

Characteristic of northern Mexico:____________________.

Major area of commercial grain farming:____________________.

Seasonal precipitation of 90 to 150 cm/year:____________________.

Characteristic of Quebec, Canada:____________________.

Thorn forest and monsoon forest:____________________.

Characteristic of southern Argentina:____________________.

Cfa and Dfa climate types:____________________.

Spodosols and permafrost, short summers:____________________.

Southern Spain, Italy, and Greece:____________________.

Variable temperatures but always warm:____________________.

Characteristic of Ireland and Wales:____________________.

Thick and continuous leaf canopy:____________________.

Uruguay and the La Plata:____________________.

Just west of the 98th meridian in the U.S.:____________________.

Just east of the 98th meridian in the U.S.:____________________.

Bare ground and xerophytic plants:____________________.

East coast of Madagascar:____________________.

West coast of Madagascar:____________________.

The bulk of Cuba:____.

Deciduous needleleaf trees:___________________________________.

The majority of Indonesia:___________________________________.

21. Characterize the present black rhino and white rhino populations of Africa. Include in your response the concept of *biodiversity* and the potential role that *biosphere reserves* might play in the future of these species.

____

____

____

____

____

____.

Sample Self-test

(Answers appear at the end of the study guide.)

1. A study of the interrelationships between organisms and their environment is called
   a) ecosystems
   b) ecology
   c) community
   d) biogeography

2. A study of the spatial aspects of ecology and the distribution of plants and animals is called
   a) ecosystems
   b) ecology
   c) community
   d) biogeography

3. Which of the following is <u>incorrect</u>
   a) habitat – specific location of an organism
   b) niche – function of a life-form within a given environment
   c) ecosystem – a self-regulating association of living plants and animals and their nonliving environment
   d) community – specific location of an organism

4. Photosynthesis involves
   a) the release of carbon dioxide in a process using sunlight
   b) the union of carbon dioxide and oxygen under the influence of sunlight
   c) important reactions within the stems and roots of plants
   d) processes that also operate at night

5. The net dry weight of organic matter that feeds a food chain is
   a) not related to net primary productivity
   b) biomass or the useful amount of chemical energy
   c) produced to a greater extent in tundra environments than in savannas
   d) measured in the text in oxygen per square meter per year

6. An example of a biotic ecosystem component is
   a) water
   b) calcium
   c) a geochemical cycle
   d) a heterotroph

7. The landscape devastated by the eruption of Mount Saint Helens
   a) is recovering through secondary succession
   b) has formed a type of pioneer community
   c) is now a climax community
   d) will never recover as is evident from an analysis of the land there

8. Which of the following is incorrect
   a) community – interacting populations of plants and animals
   b) ecosystems – interplay between communities and the physical environment
   c) life-form – outward physical properties of plants
   d) LME – an early life form in the tropics

9. A biome is
   a) an ecosystem characterized by related animal populations
   b) a large, stable terrestrial ecosystem or aquatic ecosystem
   c) the smallest local designation of a community
   d) a natural community, most of which are unaffected by human activity

10. Relative to deforestation of the rain forests,
    a) activities have not slowed, and annually exceed 6.1 million hectares
    b) has markedly slowed from previous highs due to international efforts
    c) it occurs principally for lumber and abundant specific tree species
    d) modern cutting rather than burning methods are in use

11. The biosphere reserve effort specifically involves principles of
    a) geography
    b) biology
    c) geology
    d) island biogeography

12. The pines of the southeastern United States are characteristic of which biome?
    a) tropical savanna
    b) Mediterranean shrubland
    c) temperate rain forest
    d) midlatitude broadleaf and mixed forest

13. Desertification is thought to be principally due to
  a) poor agricultural practices and overgrazing
  b) salinization
  c) global climate change
  d) political conflicts

14. A climax community is a stable, self-sustaining, and symbiotically functioning community.
  a) true
  b) false

15. The gradual enrichment of a water body with nutrients and organic materials is called eutrophication.
  a) true
  b) false

16. The final Yellowstone fire report concluded that more fire-fighting equipment and fire suppression is required to prevent further tragedy.
  a) true
  b) false

17. The media and press accurately reported the extent of the fires in Yellowstone.
  a) true
  b) false

18. The interacting population of animals defines only a community.
  a) true
  b) false

19. Gases flow into and out of leaves through the root system.
  a) true
  b) false

20. Respiration is opposite in effect to the process of photosynthesis.
  a) true
  b) false

21. Hydrogen, oxygen, and carbon comprise 99% of Earth's biomass.
  a) true
  b) false

22. A physical or chemical component that inhibits biotic activities either through excess or lack is called a limiting factor.
  a) true
  b) false

23. The tundra is found both in the extreme latitudes of North America and the former Soviet Union, and in the Southern Hemisphere.
  a) true
  b) false

24. A large, stable terrestrial ecosystem is also known as a biome.
   a) true
   b) false

25. The forests in the Amazon are also called the selva.
   a) true
   b) false

26. The floor of the equatorial rain forest is usually thick with smaller trees.
   a) true
   b) false

27. Caatinga, brigalow, and dornveld are types of chaparral associated with Mediterranean lands.
   a) true
   b) false

28. The principal division line between the tall grass and short grass prairies is the 51 cm (20 in.) annual precipitation isohyet.
   a) true
   b) false

29. The Ukraine, formerly of the Soviet Union, is an example of a midlatitude grassland.
   a) true
   b) false

Name:______________________ Class Section:______________

Date:______________ Score/Grade:__________

Earth, Humans, and the New Millennium

17

Chapter Overview

We come to the end of our journey through the pages of ELEMENTAL GEOSYSTEMS and an introduction to physical geography. A final capstone chapter is appropriate given the dynamic trends that are occurring in Earth's physical systems. Our vantage point in this course of study is physical geography at the end of one millennium and the beginning of another. The chapter asks: "Each of us might consider where we will be in life's journey on Sunday night, at 11:59 P.M., December 31, 2000, with the new millennium just one minute away?" An important personal question for each of us—goals, plans, aspirations, and the quality of life we envision!

We examine Earth through its energy, atmosphere, water, weather, climate, endogenic and exogenic systems, soils, ecosystems, and biomes, all of which leads to an examination of the planet's most abundant large animal: *Homo sapiens*.

The largest-ever gathering of nations and individuals took place at the Earth Summit in 1992. Many aspects of physical geography were at the heart of this meeting and the subject of the agreements reached. This important conference and the five agreements are summarized in FYI Report 17-1, pp. 536-38.

A note about the chapter-opening photograph by Dr. Stephen Cunha. Stephen was working in the Pamirs of Tajikistan data gathering for a national park/reserve project that is in progress. The region was depopulated during the Stalin years. Now the country is an independent member of the Commonwealth of Independent States (CIS) and people are moving back to the land of their immediate ancestors and are finding change. Disputes and unrest are daily concerns.

Dr. Cunha captured the immense beauty, developing society, and human imprint (sequent occupance) on the landscape in his exquisite photographs. His academic adventure was marked by success for his project and frightening moments of personal danger and risk of life and limb. Stephen epitomizes a geographer exploring the frontier and working the unknown using the geographic approach.

I hope this last chapter gives you something to contemplate as you leave this course. This is a time of great change in natural physical systems and therefore a time that demands a geographic perspective.

Learning Objectives

The following learning objectives help guide your reading and comprehension efforts. The operative word is in *italics*. Read and work with these carefully and note that exercise #1 asks you about five of these objectives. After reading the chapter and using this workbook, you should be able to:

1. *Resolve* an answer to Carl Sagan's question, "Who speaks for Earth?"
2. *Interpret* the significance of the quote from Marston Bates on page 530 and *relate* it to our academic activities.
3. *Explain* and *analyze* the analogy of "The Oily Bird" and *relate* this to energy consumption patterns in the United States and Canada.
4. *Overview* the *Exxon Valdez* accident and other oil spills that occurred worldwide since the accident.
5. *Review* our demands for oil and *compare* this to the remaining domestic (United States) reserves.
6. *Define* the Gaia hypothesis and *explain* the relationship attributed by this hypothesis to humans and Earth systems.
7. *Review* the impacts on the environment generated by the Persian Gulf War of 1991.
8. *Describe* examples of international cooperation that have taken place relative to Earth's environment.
9. *Review* the nuclear winter hypothesis and *relate* this to the environmental impacts of previous wars.
10. *Explain* the essential elements of the five Earth Summit agreements and *relate* them to elements covered by physical geography and in ELEMENTAL GEOSYSTEMS specifically.
11. *Discern* your place in the biosphere and your physical identity as an earthling.

Outline Headings and Glossary Review

These are the first- second-, and third-order headings that divide this chapter. The key terms and concepts that appear **boldface** in the text are listed under their appropriate heading in bold italics; these highlighted terms appear in the text glossary. A check-off box is placed next to each key term so you can mark your progress through the chapter as you define these in your reading notes or prepare note cards.

An Oily Bird

The Larger Picture

Gaia Hypothesis

The Need for International Cooperation

The Environmental Cost of Conflict

Who Speaks for Earth?

SUMMARY

Learning Activities and Critical Thinking

1. Select any five learning objectives from the list presented at the beginning of this chapter. Place the number selected in the space provided (no need to rewrite each objective). Using the following questions as guidelines only, briefly discuss your treatment of the objective.

- What did you know about the objective before you began?
- What was your plan to complete the objective?
- Which information source did you use in your learning (text, or other)?
- Were you able to complete the action stated in the objective? What did you learn?
- Are there any aspects of the objective about which you want to know more?

a)____ :____

____

____

____

____

____.

b)____ :____

____

____

____

____

____.

c)____ :____

____

____

____

____

____.

d)____ :____

____

____

____

____

____.

e)____ :____

____

____

____

____

____.

2. Beginning with the oil-contaminated western grebe shown in Figure 17-1, p. 531, outline the chain of events, linkages, and systems involved between the sick bird and you own local neighborhood shopping mall and transportation patterns.

____

____

____

____

____.

3. Overview the death toll of animals caused by the *Exxon Valdez* accident (p. 532).

____

____.

4. List the four oil spills described in the text and News Report #1, p. 534, that exceeded the Alaskan spill. Remember that approximately 10,000 oil spills occur each year across the globe as represented on the map in Figure 17-2.

____

____.

5. Given items 2, 3, and 4 just completed, and knowing the other impacts of fossil fuel combustion on the atmosphere, climate, precipitation chemistry, and land, briefly state *your point of view* and *opinion* as to what society should do either to alleviate these problems, or reduce their impacts, or to continue business as usual. Be as specific as possible about your suggested policies and position on the issues. There is no right or wrong answer here, just your answer!

____

____

____

____

____

____

____

____

____

____

____

____.

6. Review the agreements that emerged from the United Nations Conference on Environment and Development—Earth Summit of 1992 (pp. 536-38).

(a) ____

____

____.

(b) ____

____

____.

(c) ____

____

____.

(d) ____

____

____.

(e) ____

____

____.

7. What is the U. N. Commission on Sustainable Development? ______________

____

____.

8. ELEMENTAL GEOSYSTEMS presents Earth's physical and human systems as integrated and linked in complex webs of energy and material flows. The June 1991 eruption of Mount Pinatubo in the Philippines illustrates the spatial aspects of Earth's systems in a dramatic fashion. Given the text pages and sections noted below, describe the various physical phenomena triggered by the largest volcanic eruption so far this century.

(a) <u>p. 7, p. 53</u>: ____

____

____.

(b) <u>p. 78</u>: ____

____

____.

(c) p. 109: ______________________________

____

______________________________.

(d) p. 241: ______________________________

____

______________________________.

(e) p. 244: ______________________________

____

______________________________.

(f) p. 308: ______________________________

____

______________________________.

(g) p. 538: ______________________________

____

______________________________.

A final thought: the last pages of Chapter 17 demonstrate an increasing awareness of Earth and environmental issues, for in 1989, *Time* magazine named Earth the "Planet of the Year" instead of its normal practice of naming a prominent citizen as its person of the year. Each Earth Day, every April 22, is celebrated by more and more people. Local communities, governmental institutions, and corporations are overwhelmed by the willingness of the public to reduce, reuse, and recycle resources. These "representative" bodies consistently lag behind the public's growing concern for the environment and increasing desire for sustainable behavior and long-term perspectives.

Earth-system scientists are assessing the environment with increasing sophistication. Many of these new methods are mentioned throughout this text. Despite this scientific activity, measurement, and confirmation, there remain a few media personalities who have for unexplained reasons captured popular interest by declaring that none of the changes in the ozone layer, or increases in acid deposition, or the dynamics of climate change, among othe issues, are actually occurring.

The anti-science "pop" rhetoric became s intense that the American Association for th Advancement of Science (AAAS) ran a stron condemnation of the media circus. Please se Gary Taubes, "The Ozone Backlash," *Scienc* 260 June 11, 1993: 1580-83. The article detail the misrepresentation and misunderstandin by the pop-critics as they formulated their il founded and loudly stated views that every thing is all right in the stratosphere an elsewhere. The challenge to all of us especially those in the leading economies tha extract the most Earth resources, is to learn apply, and behave in a more responsibl manner than has dominated the industria revolution to date.

Carl Sagan answered his question tha opens this chapter, "Who speaks for Earth?" "We speak for Earth," he answered. We ar the earthlings. May we all perceive our spatia importance within Earth's ecosystems and d our part to maintain a life-supporting Eart into the future.

Please feel free to communicate questions, ideas, opinions, and your thoughts about the subjects we have shared in ELEMENTAL GEOSYSTEMS. I will attempt to respond, revise, correct, and update future editions of the text and this workbook. The best to you, your academic efforts, and your future—fellow Earthling.

Robert W. Christopherson
American River College
Sacramento, CA 95841

Answer Key to Self-tests

Chapter 1

1. c
2. a
3. b
4. d
5. c
6. d
7. b
8. d
9. a
10. b
11. a
12. a
13. b
14. a
15. b
16. spatial
17. open; closed; closed
18. Polaris (North Star); Southern Cross (Crux Australis)
19. Primary standard; *NIST-7*, National Institute for Standards and Technology
20. Check definitions from glossary for great circle, small circle
21. a) cylindrical, b) conic, c) planar, d) oval

Chapter 2

1. b
2. d
3. b
4. d
5. d
6. d
7. a
8. c
9. a
10. a
11. d
12. b
13. b
14. a
15. b
16. b
17. a
18. c
19. b
20. d
21. a
22. b
23. a
24. a
25. a
26. gamma, x-rays, ultraviolet wavelengths; visible light wavelengths; and infrared wavelengths.
27. geographic information system
28. b
29. b
30. b

Chapter 3

1. b
2. c
3. d
4. b
5. d
6. c

7. c
8. b
9. fresh snow; asphalt (5-10%, Moon 6-8%)
10. CO_2, methane, CFCs, H_2O vapor
11. c
12. a
13. b
14. c
15. c
16. d
17. microclimatology, including boundary layers climates
18. e
19. e
20. a
21. a
22. b
23. a
24. a
25. b
26. b
27. a
28. –273°C, –459.4°F, 0 K; 32°F (0°C, 273 K); 212°F (100°C, 373 K); United States.
29. Vostok, Antarctica, 78°S 106°E, –89°C (–129°F), July 21, 1983.
30. Al'Aziziyah, Libya, 32°N 117°W, 57°C (134°F), July 10, 1922.

Chapter 4

1. a
2. b
3. b
4. a
5. c
6. c
7. c
8. a
9. a
10. b
11. a
12. a
13. b
14. a
15. b
16. b
17. a
18. b
19. a
20. The piling up of ocean water along the western margin of each ocean basin produced by the trade winds that drive the oceans westward.

Chapter 5

1. b
2. c
3. d
4. c
5. d
6. a
7. b
8. c
9. personal response
10. d
11. a
12. d
13. a
14. d
15. b
16. a
17. b
18. b
19. b
20. a
21. a
22. a
23. b
24. hair hygrometer; sling psychrometer vapor pressure.
25. fog; 1000m (3300 ft); advection fog; evaporation fog; radiation fog.
26. a
27. b
28. a
29. b

30. a
31. a
32. a
33. b
34. windward; leeward

Completed Analysis for the Daily Weather Map–April 1, 1971

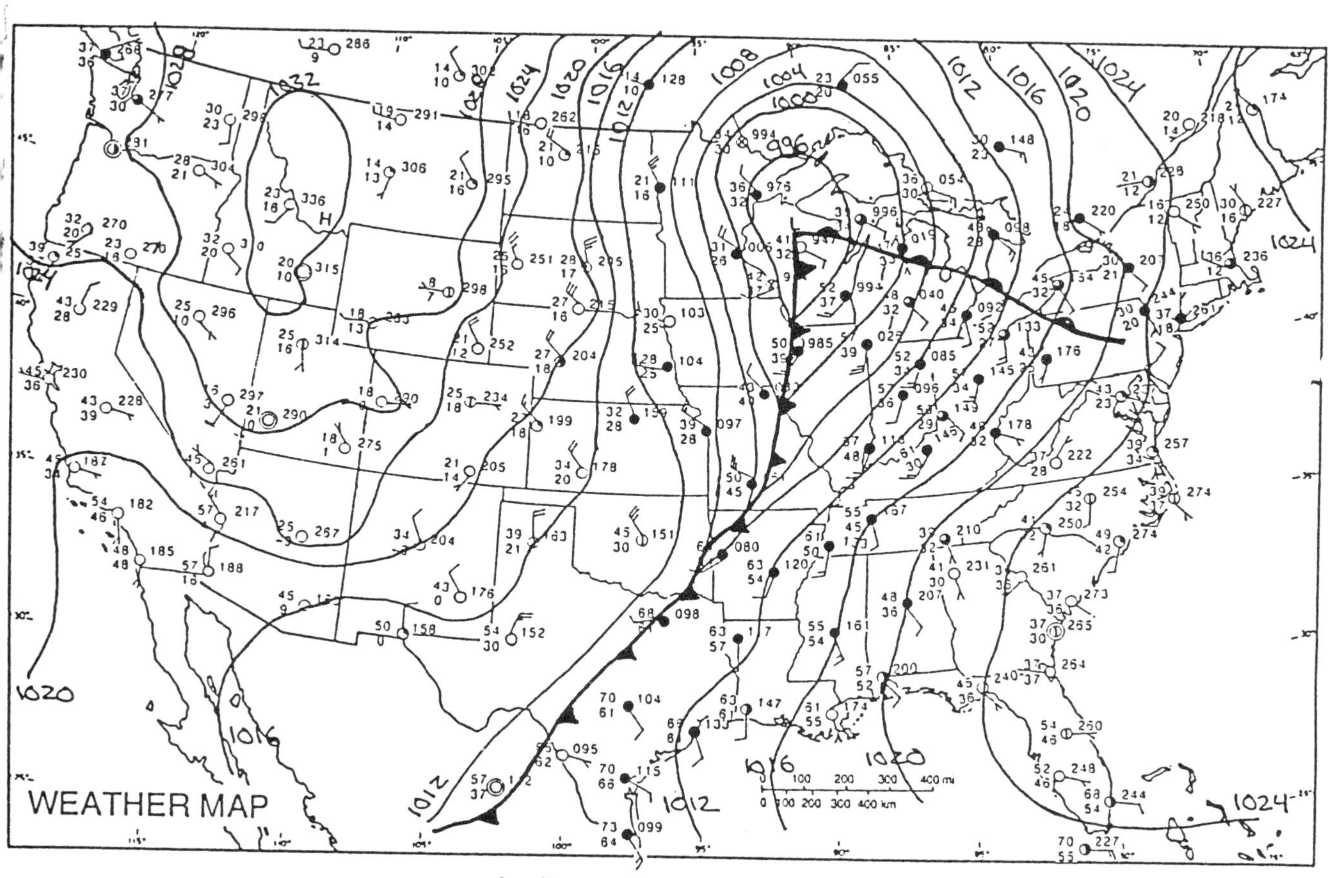

April 1, 1971, 7AM, EST

Chapter 6

1. b
2. b
3. c
4. d
5. d
6. c
7. c
8. a
9. d
10. d
11. d
12. a
13. a
14. a
15. b
16. a
17. b
18. a
19. a
20. a
21. b

Chapter 7

1. d
2. c
3. d
4. d
5. d
6. c
7. d
8. a
9. d
10. a
11. d
12. b
13. b
14. a
15. b
16. b
17. a
18. a
19. a
20. b

Chapter 8

1. c
2. d
3. b
4. d
5. b
6. d
7. a
8. d
9. b
10. a
11. b
12. b
13. b
14. a
15. b
16. a
17. a
18. a
19. uniformitarianism; present...past; catastrophism.
20. mineral; 99%; rock.

Chapter 9

1. a
2. c
3. c
4. d
5. c
6. b
7. a
8. a
9. b
10. a
11. b
12. b
13. a
14. a
15. a
16. b
17. a
18. a
19. a
20. a

Chapter 10

1. c
2. d
3. e
4. a
5. d
6. a
7. a
8. c
9. a
10. a
11. a
12. a
13. a
14. a
15. a
16. b
17. geomorphology
18. Karst topography; Krs
19. soil creep
20. scarification

Chapter 11

1. b
2. a
3. b
4. a
5. d
6. b
7. b
8. b
9. a
10. a
11. b
12. b
13. a
14. b
15. a
16. a
17. b
18. a
19. Atlantic; Gulf of Mexico; Pacific; Pacific-Bering Sea
20. parallel; rectangular; radial; annular
21. Atchafalaya River
22. 7 forms; 500 years

Chapter 12

1. c
2. c
3. a
4. c
5. c
6. b
7. d
8. d
9. b
10. b
11. a
12. b
13. a
14. a
15. a
16. a
17. b
18. a

Chapter 13

1. d
2. c
3. c
4. a
5. b
6. a
7. d
8. c
9. a
10. b
11. a
12. a
13. b
14. b
15. a
16. b
17. b
18. b

19. (See labels in Figure 1, Focus Study 13-1, text p.416; and your work here)
20. transition; translation
21. Barrier spit; bay barrier; tombolo
22. salt marshes; mangrove swamps

Chapter 14

1. c
2. a
3. e
4. b
5. c
6. d
7. d
8. b
9. a
10. a
11. a
12. b
13. b
14. b
15. a
16. a
17. b
18. a
19. a
20. a
21. arete; col; horn; tarn; paternoster.
22. Louis Agassiz (1807-1873), professor of natural history.

Chapter 15

1. c
2. b
3. a
4. a
5. c
6. d
7. c
8. d
9. a
10. a
11. b
12. a
13. a
14. b
15. b
16. b
17. a
18. a
19. a
20. a

Chapter 16

1. b
2. d
3. d
4. b
5. b
6. d
7. a
8. d
9. b
10. a
11. d
12. d
13. a
14. a
15. a
16. b
17. b
18. b
19. b
20. a
21. a
22. a
23. b
24. a
25. a
26. b
27. b
28. a
29. a